N.N. Korneev, A.F. Popov, and B.A. Krentsel'

COMPLEX ORGANOMETALLIC CATALYSTS

Translated from Russian by J. Schmorak

Israel Program for Scientific Translations
Jerusalem 1971

This book is a translation of

KOMPLEKSNYE METALLORGANICHESKIE KATALIZATORY

Izdatel'stvo "Khimiya"
Leningrad 1969

SBN 7065 1110 7

IPST Cat. No. 2228

Printed and Bound in Israel
Printed in Jerusalem by Keter Press
Binding: Wiener Bindery Ltd., Jerusalem

V/10

Table of Contents

FOREWORD

This book deals with the very topical problems of synthesis and utilization of the large group of organometallic compounds which are now being intensively studied in several countries. This is because the employment of complex organometallic catalysts makes it possible to prepare stereoregular polymers, some of which are of high industrial importance (e.g., polyolefins).

The discovery of complex organometallic catalysts was a turning point in the further development of polymer science as a whole and gave rise to new branches of organic and physical chemistry, the importance of which can hardly be overestimated.

A large number of publications on the subject are now available. They deal with the various aspects of stereospecific polymerization realized with the aid of complex catalytic systems. This notwithstanding, information on the preparation of the components of these systems, on their properties and on the analytical methods is scanty and is mostly confined to patent publications.

The authors of the present book have several years' experience in the preparation and utilization of various complex catalytic systems, and the book will thus largely fill this gap. It is the only monograph available in which the problems involved in the preparation and the study of the properties of complex catalysts, as well as the various fields of their utilization, are treated in such great detail. It will no doubt come to serve as a very useful textbook and reference book not only to chemists and physical chemists engaged in the study of high polymers, but also to research workers and technicians engaged in the various fields of organic synthesis.

The book is valuable, firstly, because it contains abundant, carefully selected material, both compiled from isolated publications, and representing the results of the authors' own work and, secondly, because it points to the pathways of future development in the chemistry of complex organometallic catalysts.

Inspection of the trends prevailing in the chemistry of complex organometallic catalysts indicates even now that their use in polymerization represents only a part of their potentialities. We may expect, for example, that catalysts of this type will find use in the fixation of atmospheric nitrogen. They are employed even now, both as the complex catalysts and as their components, in industrial synthesis of higher aliphatic alcohols, higher olefins and other industrially important organic compounds. Future research on these catalysts will no doubt result in new, unexpected applications.

The book is sure to be gratefully received by a wide circle of readers and to stimulate new studies in this rapidly developing field of organic chemistry.

V. A. Kabanov
Corresponding Member of the
Academy of Sciences of the USSR

INTRODUCTION

The chemistry of organometallic compounds has by now become an independent branch of organic chemistry, which includes a large number of compounds with metal-carbon bonds in the molecule. Studies of the structure and properties of such compounds not only led to a number of new ideas in organic chemistry, but also revealed altogether new possibilities of realization of important industrial processes. A major feature of this work is the discovery of special catalytic properties displayed by certain simple and complex organometallic, especially organoaluminum compounds.

In the middle 1950's Ziegler and his coworkers discovered a new way of polymerizing ethylene into the high-molecular polyethylene with the aid of a complex organometallic compound formed from alkylaluminum and a salt of a transition metal, e.g., titanium chloride. At the same time stereospecific polymerization of propylene was effected by Natta and his school. These discoveries were an important landmark in the history of macromolecular chemistry and determined the future trend of research concerning the preparation and study of alkylaluminum compounds and other organometallic compounds and their complexes with salts of transition metals.

The possibility of preparing stereoregular polymers with unusual technological parameters by using complex organometallic catalysts evoked a vivid interest in the theoretical study of polymerizations catalyzed by these compounds. This in turn resulted in an all-round study of the structural features and related chemical properties of the various organometallic complexes, and their use in the preparation of organic compounds other than high polymers of various types.

The increasing use of organometallic complex catalysts notwithstanding, literature information on the synthesis and properties of their constituent components is very scanty. Except for a brief section in "Linear and Stereoregular Addition Polymers" by Gaylord and Mark, all information on the subject has so far appeared in scientific periodicals and patents only. We have set ourselves the task of filling this gap; this book is a review of the methods of preparation, properties, and main fields of utilization of complex organometallic catalysts.

Of the numerous varieties of such catalysts, those formed by the reaction between alkylaluminum compounds and titanium chlorides are of the greatest practical importance. This is due to their strong catalytic effect, the ready availability of the raw materials required for their preparation and the relatively simple mode of their utilization in industrial processes.

Moreover, alkylaluminum compounds are themselves highly reactive and may be used as half-products from which valuable end products — mainly primary aliphatic alcohols — are prepared. For this reason the preparation of components of complex catalysts from alkylaluminum compounds and titanium chlorides will be described in the greatest detail. Other complex organometallic catalyst systems are treated less exhaustively, while π-allyl complexes of transition metals are not dealt with at all. These complexes have been recently studied in order to clarify the mechanism of stereoregulation during the polymerization of dienic hydrocarbons. The special features of their formation, properties and modes of utilization must form the subject of another study.

The mechanism of formation and catalytic effect of complexes of alkylaluminum compounds with transition metal compounds have been extensively studied during the past decade. Discrepancies between experimental results obtained under dissimilar conditions often make it difficult to arrive at conclusions of general validity, and the mechanism of complex formation and catalytic effect of the complex formed is not yet quite clear. These problems are outside the scope of this book, except for those which, in our view, must be discussed in order to be able to understand and critically evaluate the different methods for the preparation of both the starting substances and the complex organometallic catalysts themselves.

The scope of application of the complex organometallic catalysts and their individual constituents keeps increasing. For this reason the extensive experimental material published in scientific periodicals on the synthesis and properties of these compounds could not be reviewed comprehensively. Moreover, the personal preferences and scientific interests of the authors are necessarily reflected in the book, and some problems are dealt with in more detail than others.

The literature references given in the book are not exhaustive, but nevertheless cover the most important publications up to 1968.

All of us collaborated in the shaping and planning of the book as a whole. Chapters I and II were written by A. F. Popov, Chapters III, IV, VI and VII by N. N. Korneev, and Chapter V was written by B. A. Krentsel'. The authors wish to express their gratitude to G. B. Sakharovskaya and K. L. Makovetskii for their help in writing individual sections, and to L. L. Stotskaya and G. K. Korneeva for the editing of the experimental material.

Chapter I

ORGANOALUMINUM COMPOUNDS

Organoaluminum compounds are defined as aluminum compounds which contain at least one bond between the hydrocarbon group and the aluminum atom in the molecule. The remaining valencies of the aluminum atom may form bonds with hydrogen, halogens, an alkoxy group, an amino group, etc.

The nomenclature used in this book is that of Zhigach and Stasinevich /1/. The following are the most important appellations of the different groups of compounds:

AlR_3 — trialkylaluminum (aluminumtrialkyl)
AlR_2X — dialkylaluminum halide
$AlRX_2$ — alkylaluminum dihalide
$AlR_2X \cdot AlRX_2$ — alkylaluminum sesquihalide
AlR_2H — dialkylaluminum hydride
AlR_2OR' — dialkylaluminum alkoxide
$AlR(OR')_2$ — alkylaluminum dialkoxide
$AlR_2NR'_2$ — dialkylaluminum dialkylamide
AlR_2SR' — dialkylaluminum mercaptide
$AlR_3 \cdot R'_2O$ — trialkylaluminum etherate
$NaAlR_4$ — sodium aluminum tetraalkyl

The same rules apply to the naming of other compounds.

1. PHYSICOCHEMICAL PROPERTIES OF ALKYLALUMINUM COMPOUNDS

Alkylaluminum compounds are colorless liquids. The lower compounds (with not more than 4 carbon atoms in the radical) are spontaneously flammable in the air; alkylaluminum compounds with larger hydrocarbon groups are slowly oxidized in the air.

Most triarylaluminum compounds and their derivatives are crystalline compounds, which are not flammable in the air even when concentrated.

Owing to the high reactivity of alkylaluminum compounds the number of solvents in which they dissolve without decomposition is small. They include, first and foremost, aliphatic and aromatic hydrocarbons. Arylaluminum compounds are readily soluble in aromatic hydrocarbons and are practically insoluble in paraffins /2/. It has been recently shown that alkylaluminum compounds are soluble in ethyl chloride /3, 4/ and in carbon tetrachloride /5/*. Alkylaluminum compounds form complexes with ethers, thioethers, amines, phosphines and other compounds of similar structure.

* It should be borne in mind that carbon tetrachloride reacts vigorously, and sometimes explosively, with alkylaluminum compounds, especially so if the concentration of the compound in solution is high /159/.

All alkylaluminum compounds with a straight hydrocarbon chain known so far are dimeric. In compounds such as trimethylaluminum, triethylaluminum and tripropylaluminum, for example, the dimeric molecules are stable even in the gas phase and dissociate only above 100°C /6 — 9/. In dilute solutions trialkylaluminum compounds dissociate; thus, triethylaluminum is monomeric in 0.1% solution in benzene. However, the establishment of equilibrium after dilution is slow.

Trialkylaluminum compounds with branched hydrocarbon chains, such as triisopropylaluminum, triisobutylaluminum, trineopentylaluminum, etc., are monomeric /9, 11/, while alkylaluminum hydrides are trimeric /11/. Arylaluminum compounds (e.g., triphenylaluminum) are partly associated /9/.

The strong tendency of alkylaluminum compounds to associate is the result of the electron deficiency at the aluminum atom in compounds in which aluminum has the coordination number 3. This deficiency has a significant effect on the physical and chemical properties of organoaluminum compounds and is responsible for their association by way of so-called "electron-deficient" bonds /12/ or "semi-bonds" /13/.

It was shown by Hoffman /11/ that the first carbon atoms in the alkyl groups of alkylaluminum compounds are the binding atoms, or, as they are more usually called, the bridge-forming atoms:

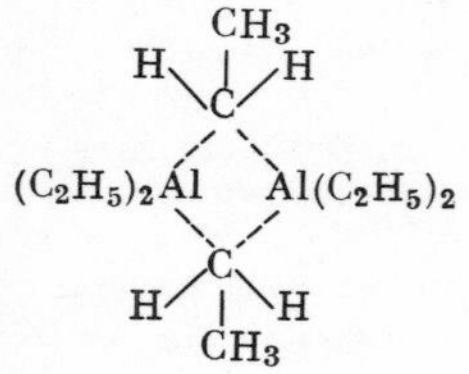

CH_3
H   H
C
$(C_2H_5)_2Al$ $Al(C_2H_5)_2$
C
H   H
CH_3

Triisobutylaluminum and other alkylaluminum compounds with branched chains are monomeric, since association is prevented by steric hindrances; dialkylaluminum hydrides are cyclic trimers, since groups capable of associating would be left at the chain ends, were the molecule linear. In this case the association is effected via the hydrogen directly bound to aluminum:

AlR_2
H   H
R_2Al AlR_2
H

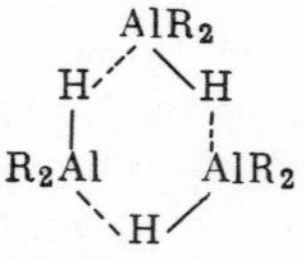

Diisobutylaluminum hydride also has a hydrogen bridge structure in the form of a six-membered ring (degree of association 2.4) which is another illustration of the fact that the tendency to associate largely depends on the surroundings of the bridge atom: in alkylaluminum hydrides the bridge is constituted only by hydrogen atoms directly bound to aluminum.

X-ray studies /13, 14/, IR spectra /15/ and NMR spectra /16/ established the following symmetrical formula for the dimer of trimethylaluminum:

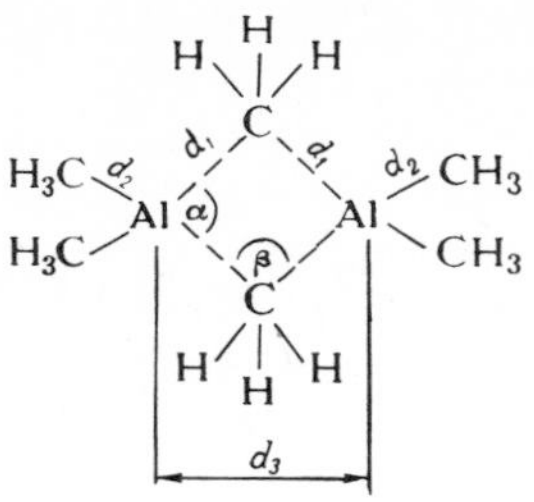

$d_1 = 2.24$ Å
$d_2 = 1.89$ Å
$d_3 = 2.55$ Å

The valency angles α at the aluminum atoms and β at the bridge carbon atoms are 110 and 70° respectively /73/.

Cleavage of an associated organoaluminum compound into its constituents may be attained by the addition of electron donors. Etherates, thioetherates and aminates of trialkylaluminum compounds are monomolecular /17/, stable compounds which distil without decomposition /18a/. Takeda et al. prepared 15 complexes of organoaluminum compounds with ethers, which distil over in vacuo at 41 — 165°C /18b/. It was found that the electron-accepting capacity of alkylaluminum compounds increases in the sequence $AlR_3 < AlR_2X < AlRX_2$, where R is methyl or ethyl and X is a halogen. It may be pointed out in this connection that alkylaluminum fluorides, unlike other alkylaluminum halides, do not form stable etherates /18/.

Certain properties of alkylaluminum compounds change as a result of the addition of ethers, amines or similar compounds. Thus, for instance, owing to the strong polarization /19, 20/, the addition products have a large dipole moment (4 — 6 D), while neither the associated nor the monomolecular trialkylaluminum compounds have significant dipole moments.

Complex formation with electron donors is also accompanied by characteristic changes in the electrical conductivity of the solutions. Highly concentrated trialkylaluminum compounds and dialkylaluminum hydrides have electrical conductivities of the order of 10^{-10} ohm^{-1} cm^{-1}; electrical conductivities of trialkylaluminum etherates and aminates are $20 \cdot 10^{-7}$ ohm^{-1} cm^{-1} /21/. It is also interesting to note that the electrical conductivity of a mixture of triethylaluminum with diethylaluminum hydride in a molar ratio of 1 to 1, which did not contain a solvent, was $1.8 \cdot 10^{-8}$ ohm^{-1} cm^{-1} /22/. The difference between the polarization and between the electrical conductivities of simple and complex alkylaluminum compounds was utilized by Bonitz and Graevskii /21 — 23/ in the determination of these compounds by conductometric and potentiometric titrations.

Trialkylaluminum compounds are unstable at elevated temperatures. Even lower trialkylaluminum compounds distil over with decomposition under atmospheric pressure. Triethylaluminum redistilled at 80 — 100°C in vacuo without any special precautions contains up to 5% diethylaluminum hydride. The decomposition is not complete below 200°C. Trialkylaluminum compounds with branched chain groups deserve special mention. Thus, for instance, triisobutylaluminum is 50% decomposed at 100°C in one hour, with evolution of isobutylene. The dependence of the time $Q_{1/2}$ required for half the material to become decomposed under 4 mm Hg pressure on temperature T may be written as follows /24/:

$$\lg Q_{1/2} = \frac{4918}{T} - 13.222$$

The pyrolysis of trimethylaluminum was studied in fair detail as early as 1946 /25/. Trimethylaluminum decomposes above 300°C, with the evolution of methane, ethane and hydrogen. The solid residue from pyrolysis consisted of aluminum, aluminum carbide and polymeric products. The author reports the kinetic results of thermal decomposition of trimethylaluminum. The most extensive studies of pyrolysis of trialkylaluminum compounds were conducted by Ziegler et al. /26/. These workers showed that the pyrolysis of trialkylaluminum compounds involves not only the decomposition according to the well-known scheme

$$R_2AlR \longrightarrow R_2AlH + CH_2{=}CH{-}R'$$
$$R_2AlH \longrightarrow RAlH_2 + CH_2{=}CH{-}R'$$
$$3RAlH_2 \longrightarrow 2AlH_3 + 3CH_2{=}CH{-}R' + Al$$
$$AlH_3 \longrightarrow Al + 1^1/_2H_2$$

but also side reactions which result in the formation of a large number of side compounds, e.g.:

$$Al[CH_2{-}CH(CH_3)_2]_3 \longrightarrow Al(CH_3)_3 + 3CH_3{-}CH{=}CH_2$$
$$Al(CH_3)_3 \longrightarrow CH_4 + [CH_3{-}\underset{|}{Al}{-}\underset{|}{CH_2}]_n$$
$$2Al(CH_3)_3 \longrightarrow CH_4 + (CH_3)_2Al{-}CH_2{-}Al(CH_3)_2$$
$$2nAl(CH_3)_3 \longrightarrow [Al_2(CH_2)_3]_n + 3nCH_4$$
$$Al(CH_3)_3 \longrightarrow (CH{\equiv}Al)_n + 2nCH_4\text{ , etc.}$$

Similar side processes also take place during the pyrolysis of other trialkylaluminum compounds at 200 — 300°C. Larikov et al. /27/ studied the thermal decomposition of triethylaluminum and triisobutylaluminum in the liquid phase at a constant volume and under constant pressure. He confirmed that between 50 and 180°C the main decomposition reaction is the dissociation into dialkylaluminum hydride and olefin. Between 180 and 300°C triisobutylaluminum decomposes into hydrogen, aluminum and isobutylene, while triethylaluminum decomposes to form a complex mixture of alkylaluminum compounds and hydrocarbons. The respective temperatures of incipient thermal decomposition of triethylaluminum and triisobutylaluminum in an enclosed space are 150 and 50°C. These experimental data served to calculate the heats of dissociation and to determine the equilibrium constants.

Trialkylaluminum compounds with branched chain groups (especially secondary and tertiary) isomerize at elevated (110 — 130°C) temperatures to give alkylaluminum compounds with straight chain groups /28/. Conversions of this type were subsequently noted by Natta et al. when tri(1-phenylethyl)aluminum was heated at 80°C in the presence of colloidal nickel; the result was isomerization to tri(2-phenylethyl)aluminum /29/.

Substituted alkylaluminum and arylaluminum compounds

Substituted alkylaluminum and arylaluminum compounds of the general formula $R_{3-n}AlX_n$, in which X is a halogen, $-R$, $-OR$, $-SR$, or $-C\equiv CH$, etc., are mostly viscous liquids or crystalline substances. Thus, for instance, dialkylaluminum halides are colorless liquids which are readily oxidized in the air. The lower members of the series burst into flame in the air. Alkylaluminum dihalides are crystalline solids under normal conditions.

Dialkylaluminum alkoxides are not spontaneously flammable, but they fume in the air. Alkylaluminum amines are even more inert. Most of these compounds are readily soluble in hydrocarbons, both aliphatic and aromatic, are fairly stable to heating, and distil over without decomposition in vacuo up to 140 — 160°C.

Substituted alkylaluminum compounds are associated, similarly to trialkylaluminum compounds. Many of them are known to form dimers and trimers /11, 17, 17a/:

$$R_2Al\langle Cl_2 \rangle AlR_2 \qquad R_2Al\langle (OR)_2 \rangle AlR_2 \qquad R_2Al\langle (NR_2)_2 \rangle AlR_2$$

$$[R_2Al(OR)]_3 \qquad [R_2AlPR_2]_3$$

The bridges in this case consist of oxygen, halogen, nitrogen and phosphorus atoms.

Alkylaluminum fluorides are so strongly associated that they are converted into actual polymers. Thus, diethylaluminum fluoride distils over at about 200°C; at this temperature it is a mobile liquid, but solidifies on cooling to form a vitreous substance /18/.

Substituted alkylaluminum products, like AlR_3, form complexes with amines, ethers and other electron-donating compounds. The associated molecules dissociate into monomers at the same time: dialkylaluminum halide etherates and also trialkyl aminates and thioetherates of these compounds are monomeric /11/. Another interesting variety of organoaluminum donor-acceptor complexes are the alkyl compounds of aluminum with electron-donating substituents in the alkyl group, of the structure:

$$R_2Al\langle(CH_2)_n \leftarrow X \rangle \quad \text{where } X = RO,\ RS, R_2N$$

Such inner complex compounds were prepared by Zakharkin and Savina /30/. The determination of the molecular weight of these compounds showed that those with $n = 3$ and $n = 4$ are in fact inner complexes. In compounds with $n = 5$, in which the formation of seven-membered rings is possible, the experimentally found molecular weight was higher than the calculated value and for this reason these compounds must not be considered as purely inner-complex compounds; some of the complex bonds are intermolecular.

The physicochemical constants of organoaluminum compounds reported by different workers have been assembled by Zhigach and Stasinevich /1/; newly determined parameters will be found in /22/ and in the book by Rochow, Hurd and Lewis /31/. Table 1 gives the main physicochemical constants of alkylaluminum compounds which are most frequently employed in the different chemical processes.

2. CHEMICAL PROPERTIES OF ORGANOALUMINUM COMPOUNDS

We have mentioned that alkylaluminum compounds are very reactive. A major reason for it is the fact that compounds of the types R_3Al, R_2AlX and $RAlX_2$ are electron-deficient. Accordingly, alkylaluminum compounds form complexes with nucleophilic reagents such as ethers, tertiary amines, thioethers, etc. These compounds display many properties typical of Grignard reagents.

The more important reactions of organoaluminum compounds are given below.

Reactions with elements of Group I and their compounds

Trialkylaluminum compounds react with hydrogen at elevated temperatures and pressures to form dialkylaluminum hydrides /41/:

$$AlR_3 + H_2 \longrightarrow AlR_2H + RH$$

They react with alkali metals to form the metal aluminum tetraalkyl /38/:

$$4Al(C_2H_5)_3 + 3Na \longrightarrow 3NaAl(C_2H_5)_4 + Al$$

Reactions between a metal aluminum tetraalkyl and an alkali metal amalgam are of the exchange type /42/:

$$NaAl(C_2H_5)_4 + K \longrightarrow KAl(C_2H_5)_4 + Na$$

Trialkylaluminum compounds, dialkylaluminum halides and alkylalkoxyaluminum halides react with alkali metal hydrides as follows /43 — 46/:

$$(C_2H_5)_3Al + NaH \longrightarrow Na[Al(C_2H_5)_3H]$$
$$(C_2H_5)_2AlCl + LiH \longrightarrow (C_2H_5)_2AlH + LiCl$$
$$(C_2H_5)_2AlCl + NaH \longrightarrow (C_2H_5)_2AlH + NaCl$$
$$(C_3H_7)_2AlCl + 2LiH \longrightarrow Li[Al(C_3H_7)_2H_2] + LiCl$$
$$Cl(C_2H_5)AlOC_2H_5 + NaH \longrightarrow Na[Al(C_2H_5)Cl(OC_2H_5)H]$$

When ethylaluminum dichloride was made to react with lithium and sodium hydrides, a mixture of products was obtained which contained $(C_2H_5)_2AlH$, $(C_2H_5)AlH_2$ and $NaAlH_4$.

Complex compounds are formed with alkali metal salts /26, 47/, according to the following equations:

$$MX + AlR_3 \longrightarrow M[AlR_3X]$$

where M=Na, K, Rb, Cs; X=F, Cl, Br, CN, CNS

$$MX + 2AlR_3 \longrightarrow MX \cdot 2AlR_3$$

where X=F, Cl, CN.

Lithium, sodium and potassium hydrides displace alkali metal cyanides and fluorides from their complexes with trialkylaluminum compounds /48/:

$$NaCN \cdot 2Al(C_2H_5)_3 + 2NaH \longrightarrow 2Na[Al(C_2H_5)_3H] + NaCN$$

Trialkylaluminum compounds form metal aluminum tetraalkyls with alkali metal alkyls /49, 50/:

$$Al(CH_3)_3 + LiCH_3 \longrightarrow Li[Al(CH_3)_4]$$
$$Al(C_2H_5)_3 + NaC_2H_5 \longrightarrow Na[Al(C_2H_5)_4]$$

Reactions with elements of Group II and their compounds

Freshly prepared magnesium powder catalyzes the reaction between alkylaluminum compounds and hydrogen /51/; the reaction results in the formation of dialkylaluminum hydrides.

The action of metallic magnesium on alkylaluminum halides results in dehalogenation /52 — 54/:

$$2RAlCl_2 \cdot R_2AlCl + 1^1/_2Mg \longrightarrow 3AlR_2Cl + 1^1/_2MgCl_2 + Al$$

Alkylaluminum compounds react with mercury, zinc, cadmium and beryllium halides to form the corresponding metal dialkyls in high yields /55 — 62/:

$$ZnX_2 + 2AlR_3 \longrightarrow ZnR_2 + 2AlR_2X$$
$$CdX_2 + 2AlR_3 \longrightarrow CdR_2 + 2AlR_2X$$
$$3HgCl_2 + 2AlR_3 + 2NaCl \longrightarrow 3HgR_2 + 2NaAlCl_4$$
$$BeCl_2 + 2Al(C_2H_5)_3 \longrightarrow Be(C_2H_5)_2 + 2Al(C_2H_5)_2Cl$$

Dimethylaluminum hydride reacts with dimethylberyllium to form a mixture of trimethylaluminum and beryllium hydride /63/:

$$2(CH_3)_2AlH + (CH_3)_2Be \longrightarrow 2Al(CH_3)_3 + BeH_2$$

Diethylaluminum hydride reacts with diethylmercury to form triethylaluminum, ethane, hydrogen and mercury, most probably according to the following reaction scheme /64/:

$$(C_2H_5)_2AlH + (C_2H_5)_2Hg \longrightarrow (C_2H_5)_3Al + HHgC_2H_5$$
$$HHgC_2H_5 \longrightarrow C_2H_6 + Hg$$

$$HHgC_2H_5 + (C_2H_5)_2AlH \longrightarrow (C_2H_5)_3Al + HgH_2$$
$$HgH_2 \longrightarrow Hg + H_2$$

Reaction with diethylmagnesium yielded magnesium hydride and triethylaluminum:

$$2(C_2H_5)_2AlH + Mg(C_2H_5)_2 \longrightarrow 2\,(C_2H_5)_3Al + MgH_2$$

Reaction between triethylaluminum and ethylmagnesium bromide yields magnesium dialuminum octaethyl /65/:

$$2Al(C_2H_5)_3 + 2C_2H_5MgBr \longrightarrow Mg[Al(C_2H_5)_4]_2 + MgBr_2$$

which is decomposed with the evolution of triethylaluminum and magnesium aluminum pentaethyl:

$$Mg[Al(C_2H_5)_4]_2 \longrightarrow Al(C_2H_5)_3 + (C_2H_5)_2Mg \cdot Al(C_2H_5)_3$$

The magnesium derivative of δ-chlorobutylethyl ether reacts with diethylaluminum iodide with the formation of the chelate compound δ-ethoxybutyldiethylaluminum /66/:

$$(C_2H_5)_2Al{-}I + ClMgCH_2{-}CH_2{-}CH_2{-}H_2C{-}O{-}C_2H_5 \longrightarrow (C_2H_5)_2Al\overset{\longleftarrow}{}\text{[}CH_2{-}CH_2{-}CH_2{-}H_2C{-}O(C_2H_5)\text{]} + MgICl$$

γ-Diethylaminopropyldiethylaluminum

$$(C_2H_5)_2Al{-}CH_2{-}CH_2{-}CH_2{-}N(C_2H_5)_2 \quad (N \rightarrow Al)$$

and γ-ethylmercaptodiethylaluminum

$$(C_2H_5)_2Al{-}CH_2{-}CH_2{-}CH_2{-}S{-}C_2H_5 \quad (S \rightarrow Al)$$

have been obtained in a similar manner.

Reactions with elements of Group III and their compounds

Metal-aluminum tetraalkyls — $NaAlR_4$ and $Na[AlR_3]OR$ — react with metallic aluminum in the presence of mercury with formation of alkylaluminum compounds /67/:

$$3NaAlR_4 + Al + 3Hg \longrightarrow 4AlR_3 + 3HgNa$$

$$3Na[AlR_3]OR + Al + 3Hg \longrightarrow 3AlR_2OR + AlR_3 + 3HgNa$$

Reactions between alkylaluminum compounds and aluminum halides /68 — 73/ are known to yield alkylaluminum halides:

$$2AlR_3 + AlX_3 \longrightarrow 3AlR_2X$$

$$AlR_3 + 2AlX_3 \longrightarrow 3AlRX_2$$

$$Al_2R_3X_3 + AlX_3 \longrightarrow 3AlRX_2$$

where $R = CH_3$, C_2H_5, n-C_3H_7, C_6H_5, $C_6H_4CH_3$; $X = Cl$, Br, I, F.

Alkylaluminum compounds enter radical exchange reactions with organoboron compounds /74, 75/:

$$AlR_3 + BR'_3 \rightleftarrows AlR_2R' + BR'_2R$$

Trialkylboron compounds are formed by the reaction between alkylaluminum compounds and boron halides /76, 77/, boric acid esters /78/ and boric anhydride /79/.

It was shown by Zakharkin and Gavrilenko /81/ that the reduction of boron halides by sodium hydride to sodium borohydride readily takes place in the presence of trialkylaluminum compounds, which give soluble complexes with sodium hydride:

$$AlR_3 + NaH \longrightarrow Na[AlR_3]H$$

These complexes rapidly reduce boron halides to sodium borohydride:

$$3Na[AlR_3]H + BCl_3 + NaH \longrightarrow NaBH_4 + 3AlR_3 + 3NaCl$$

Ziegler et al. /82/ prepared higher organoboron compounds by reacting ethylene with lower trialkylboron compounds in the presence of alkylaluminum compounds:

$$B(C_2H_5)_3 + 3nC_2H_4 \xrightarrow{AlR_3} B[(C_2H_4)_nC_2H_5]_3$$

Alkylgallium and alkylthallium compounds have been prepared from trialkylaluminum compounds and gallium and thallium halides /61, 80/.

Reactions with elements of Group IV and their compounds

Reactions between alkanes and organoaluminum compounds have not been described in the literature. Ziegler /83/ reported an exchange reaction between benzene and sodium aluminum tetraalkyl in the presence of sodium alcoholate as catalyst:

$$NaAl(C_2H_5)_4 + 4C_6H_6 \xrightarrow{NaOC_2H_5} NaAl(C_6H_5)_4 + 4C_2H_6$$

Nicolescu et al. /84/ studied the reaction between aromatic hydrocarbons and organoaluminum compounds. They showed that ethylaluminum sesquibromide efficiently catalyzes the alkylation reaction of aromatic hydrocarbons by cyclohexene. Alkylaluminum halides can probably also react with naphthalene /85/:

$$C_{10}H_8 + R_2AlX \longrightarrow C_{10}H_8(R)(AlRX) \xrightarrow{C_2H_5OH} C_{10}H_9R$$

Numerous examples are known of reactions between organoaluminum compounds and olefins /22, 26, 86 — 97/; these may be subdivided into two groups:

e x t e n s i o n of alkyl chains

$$Al(C_2H_5)_3 + (n+m+p)C_2H_4 \longrightarrow Al\begin{cases}(C_2H_4)_nC_2H_5\\(C_2H_4)_mC_2H_5\\(C_2H_4)_pC_2H_5\end{cases}$$

and d i s p l a c e m e n t reactions

$$Al(C_2H_4{-}R)_3 + 3C_2H_4 \longrightarrow Al(C_2H_5)_3 + 3R{-}CH{=}CH_2$$

These reactions are extensively utilized in industry in the preparation of higher olefins, higher alcohols and various organoaluminum compounds.

The composition of the final reaction products will depend on the structures of the initial alkylaluminum compound and olefin. Trialkylaluminum compounds which are not branched at the α-carbon atom of the hydrocarbon radicals add on to α-olefins and cyclic olefins /98, 99/. Dialkylaluminum hydride and styrene react in two ways /100/:

$$R_2AlH + CH_2{=}CHC_6H_5 \longrightarrow R_2AlCH_2CH_2C_6H_5$$

$$R_2AlH + \underset{\underset{CH_2}{\|}}{CH}{-}C_6H_5 \longrightarrow R_2Al\underset{\underset{CH_3}{|}}{CH}C_6H_5$$

When triethylaluminum is made to act on allyl compounds of the type $CH_2 = CH - CH_2X$ (where X = OR, SR, NR_2), the C — X bond is broken with liberation of the olefin and formation of $(C_2H_5)_2AlX$ /30/:

$$(C_2H_5)_2AlC_2H_5 + CH_2{=}CH{-}CH_2X \longrightarrow CH_2{=}CH{-}CH_2{-}C_2H_5 + (C_2H_5)_2AlX$$

Diisobutylaluminum hydride adds on to the double bond of olefins of this type:

$$(\text{iso-}C_4H_9)_2AlH + CH_2{=}CH{-}CH_2X \longrightarrow (\text{iso-}C_4H_9)_2AlCH_2CH_2CH_2X$$

Vinyl butyl ether reacts with diisobutylaluminum hydride with formation of ethylene and diisobutylbutoxyaluminum:

$$(\text{iso-}C_4H_9)_2AlH + CH_2{=}CH{-}OC_4H_9 \longrightarrow CH_2{=}CH_2 + (\text{iso-}C_4H_9)_2AlOC_4H_9$$

Of practical importance is the reaction of isomerization of trialkylaluminum compounds /28, 103/

$$Al(\text{iso-}C_nH_{2n+1})_3 \longrightarrow Al(\text{n-}C_nH_{2n+1})_3$$

by which olefins with double bonds not situated at the end of the molecule can be used in the synthesis of organoaluminum compounds with n-alkyl groups, α-olefins and α-alcohols.

Asinger et al. /4/ showed that the double bond of an olefin may undergo a catalytic shift from the middle of the chain to the α-position in the presence of a transition metal salt and alkylaluminum compound /104/. The reaction proceeds readily with a high yield.

Reactions between dialkylaluminum hydrides and trialkylaluminum compounds and acetylene yield alkenyl derivatives of aluminum:

$$R_2AlH + CH{\equiv}CH \longrightarrow R_2AlCH{=}CH_2$$

Alkynylaluminum compounds are prepared as follows /105, 106/:

$$Al(C_2H_5)_3 + HC{\equiv}CR \longrightarrow (C_2H_5)_2AlC{\equiv}CR + C_2H_6$$

$$R_2AlX + MeC{\equiv}CH \longrightarrow R_2AlC{\equiv}CH + MeX$$

Reactions between dialkylaluminum hydrides and alkynylaluminum compounds may yield bifunctional or trifunctional organoaluminum compounds /83/:

$$R_2Al{-}C{\equiv}CH \xrightarrow{R_2AlH} (R_2Al)_2C{=}CH_2 \xrightarrow{R_2AlH} (R_2Al)_3C{-}CH_3$$

which disproportionate with the separation of a trialkylaluminum compound and formation of a cyclic organoaluminum compound, the so-called hexaaluminumadamantane.

Electrolysis of complex compounds of aluminum alkyls in vats with lead or tin electrodes yields organic compounds of lead and tin /107/:

$$Pb + 4NaAlR_4 \longrightarrow PbR_4 + 4AlR_3 + 4Na$$

Many alkylation reactions of tin, lead, silicon and germanium by alkylaluminum halides and trialkylaluminum compounds have been described /59, 61, 62, 108, 109/:

$$3SnCl_4 + 4AlR_3 + 4NaCl \longrightarrow 3SnR_4 + 4NaAlCl_4$$

$$PbX_2 + AlR_3 + RX \longrightarrow PbR_4 + AlX_3$$

$$Pb + 2RX + 2NaAlR_4 \longrightarrow PbR_4 + 2AlR_3 + 2NaX$$

$$3SiCl_4 + 4AlR_3 \longrightarrow 3SiR_4 + 4AlCl_3$$

$$3GeCl_4 + 4AlR_3 \longrightarrow 3GeR_4 + 4AlCl_3$$

$$SiF_4 + AlR_3 \longrightarrow SiF_3R + AlR_2F$$

The reaction between organoaluminum compounds and titanium compounds is complicated, and the composition of the reaction products depends on the experimental conditions.

Despite the low heat-stability of organotitanium compounds, alkyltitanium trichloride is known to be one of the products of the reaction between $TiCl_4$ and alkylaluminum compounds conducted under mild conditions. Details on the course of the reaction and on the products of reaction between alkylaluminum compounds and salts of other transition metals will be found in Chapter III.

No literature references to the reaction between Group IV metals and alkylaluminum compounds are available. We may mention in this connection the report of Ziegler /110, 111/ that thermal decomposition of trialkylaluminum compounds is catalyzed by metallic titanium.

Reactions with elements of Group V and their compounds

Trialkylaluminum compounds react with ammonia and with secondary amines to form unstable complexes which split off an alkane /17, 26/:

$$AlR_3 + NH_3 \longrightarrow AlR_3 \cdot NH_3 \longrightarrow AlR_2 \cdot NH_2 + RH$$

$$AlR_3 + NHR_2 \longrightarrow AlR_3 \cdot NHR_2 \longrightarrow AlR_2 \cdot NR_2 + RH$$

Tertiary amines react to form low-melting complexes, which are sparingly soluble in organic solvents /26, 112/.

Reaction between trimethylaluminum and hydrazine yielded the crystalline, symmetrical bis(dimethylaluminum)hydrazine, which is very sensitive to impact /113/:

$$2Al(CH_3)_3 + NH_2NH_2 \longrightarrow (CH_3)_2Al\underset{\substack{|\\H}}{N}-\underset{\substack{|\\H}}{N}Al(CH_3)_2 + 2CH_4$$

When as-dimethylhydrazine was employed, the crystalline dimethylaluminumdimethylhydrazine (m. p. 77 — 78.5°C) was obtained; the compound could be distilled in vacuo /114/:

$$2Al(CH_3)_3 + 2H_2NN(CH_3)_2 \longrightarrow 2(CH_3)_2AlNHN(CH_3)_2 + 2CH_4$$

Tetramethylhydrazine yielded the complex compound

$$Al(CH_3)_3 + (CH_3)_2NN(CH_3)_2 \longrightarrow (CH_3)_3Al \cdot N(CH_3)_2N(CH_3)_2$$

Trimethylaluminum reacts with tetramethyltetrazene as follows /115/:

$$2Al(CH_3)_3 + (CH_3)_2NN{=}NN(CH_3)_2 \longrightarrow 2(CH_3)_2AlN(CH_3)_2 + N_2 + C_2H_6$$

$$2Al(CH_3)_3 + 3(CH_3)_2NN{=}NN(CH_3)_2 \longrightarrow 2Al[N(CH_3)_2]_3 + 3N_2 + 3C_2H_6$$

Alkylaminosilanes, arylaminosilanes and alkylsilazanes react with trialkylaluminum compounds according to the following scheme:

$$-\overset{|}{\underset{|}{Si}}-\overset{|}{N}-H + AlR_3 \longrightarrow -\overset{|}{\underset{|}{Si}}-\overset{|}{N}-\overset{R}{\overset{|}{Al}}-R + RH$$

$$-\overset{|}{\underset{|}{Si}}-\overset{|}{N}-\overset{R}{\overset{|}{Al}}-R + H-\overset{|}{N}-\overset{|}{\underset{|}{Si}}- \longrightarrow -\overset{|}{\underset{|}{Si}}-\overset{|}{N}-\overset{R}{\overset{|}{Al}}-\overset{|}{N}-\overset{|}{\underset{|}{Si}}- + RH$$

The reaction is exothermal. All alkylaluminum silylamides so far prepared are high-boiling, clear liquids, which are soluble in paraffinic hydrocarbons /116 — 118/. Some alkylaluminum silylamides in combination with $TiCl_4$ are effective catalysts of polymerization of olefins /119 — 121/.

Reaction between dialkylaluminum hydrides and nitriles or substituted acid amides results in their reduction to aldehydes and amines. In the latter case tetraalkylalumoxan is formed as a side product /122/:

$$R'CN + AlR_2H \longrightarrow R'CH{=}NAlR_2 \xrightarrow{\text{hydrolysis}} R'CHO$$

$$R'CONR''R''' + AlR_2H \longrightarrow R'\overset{O-Al-R_2}{\overset{|}{C}}H - N - R''R''' \xrightarrow{\text{hydrolysis}} R'CHO$$

$$\overset{OAlR_2}{RCH}-NR''R''' + AlR_2H \longrightarrow R'CH_2NR''R''' + R_2AlOAlR_2$$

The reduction of nitrous oxides by triethylaluminum etherate yields ethylnitrosohydroxylamine /50/.

Trialkylaluminum compounds react with PCl_3 to yield monoalkyldichlorophosphines, dialkylchlorophosphines and trialkylphosphines /123/.

Trialkylaluminum compounds react with dialkylphosphines at room temperature to form stable complex compounds:

$$Al(CH_3)_3 + (CH_3)_2PH \longrightarrow Al(CH_3)_3 \cdot PH(CH_3)_2$$

At elevated temperatures the reaction products have an Al — P bond:

$$Al(CH_3)_3 + (CH_3)_2PH \longrightarrow (CH_3)_2AlP(CH_3)_2 + CH_4$$

Trialkylphosphines react with the formation of complex compounds:

$$Al(CH_3)_3 + P(CH_3)_3 \longrightarrow Al(CH_3)_3 \cdot P(CH_3)_3$$

Reaction between equimolar amounts of triisobutylaluminum and triethylphosphite at room temperature yields the adduct (I), from which triethylphosphite may be regenerated by alcoholysis:

$$P(OC_2H_5)_3 + Al(\text{iso-}C_4H_9)_3 \longrightarrow \underset{(I)}{(C_2H_5O)_3P \cdot Al(\text{iso-}C_4H_9)_3}$$

$$(C_2H_5O)_3P \cdot Al(\text{iso-}C_4H_9)_3 \xrightarrow{ROH} P(OC_2H_5)_3 + Al(OR)_3 + 3\ \text{iso-}C_4H_{10}$$

At 80°C compound (I) is exothermally converted to the complex of diethyl ethylphosphonate (II). The ester may be isolated by alcoholysis of complex (II) /124/:

$$(C_2H_5O)_3P \cdot \underset{I}{Al}(\text{iso-}C_4H_9)_3 \longrightarrow C_2H_5\overset{O}{\overset{\|}{P}}(OC_2H_5)_2 \cdot \underset{(II)}{Al}(\text{iso-}C_4H_9)_3$$

$$C_2H_5\underset{O}{\underset{\|}{P}}(OC_2H_5)_2 \cdot Al(\text{iso-}C_4H_9)_3 \xrightarrow{ROH} C_2H_5\underset{O}{\underset{\|}{P}}(OC_2H_5)_2 + Al(OR)_3 + 3\ \text{iso-}C_4H_{10}$$

Arsenic, antimony and bismuth halides react with alkylaluminum compounds to form the corresponding organic derivatives /59, 61, 62/:

$$AsCl_3 + AlR_3 + NaCl \longrightarrow AsR_3 + NaAlCl_4$$

$$SbCl_3 + 3R_2AlCl + 3NaCl \longrightarrow SbR_3 + 3Na[AlRCl_2]Cl$$

$$SbF_3 + Al(\text{iso-}C_4H_9)_3 \longrightarrow Sb(\text{iso-}C_4H_9)_3 + AlF_3$$

Reactions with elements of Group VI and their compounds

Organoaluminum compounds are readily oxidized by oxygen. Lower alkylaluminum compounds are spontaneously flammable in the air. If the reaction with oxygen is carried out under controlled conditions, a stepwise synthesis of alkylaluminum alkoxides can be performed:

$$AlR_3 \xrightarrow{1/_2O_2} AlR_2OR \xrightarrow{1/_2O_2} AlR(OR)_2 \xrightarrow{1/_2O_2} Al(OR)_3$$

The first two alkyl groups are readily oxidized by atmospheric oxygen. The third group must be oxidized by oxygen in a concentrated form. The oxidation of trialkylaluminum compounds is widely employed in the preparation of various alcohols /1, 26, 86/.

It would appear that the primary reaction products are peroxy compounds $AlR_3 + O_2 \rightarrow R_2Al-OOR$, which are hydrolyzed to give the corresponding hydroperoxides. The existence of peroxides has been experimentally confirmed at low temperatures /125, 126/, but not at conventional temperatures /126/. Razuvaev et al. /127, 128/ prepared diethoxyaluminumperoxycumyl $(C_2H_5O)AlOOC_9H_{11}$ by reacting cumyl hydroperoxide with diethoxyethylaluminum. Reaction between benzoyl peroxide and triethylaluminum yielded diethylaluminum benzoate and ethyl benzoate /129/:

$$(C_6H_5COO)_2 + Al(C_2H_5)_3 \longrightarrow C_6H_5COOAl(C_2H_5)_2 + C_6H_5COOC_2H_5$$

Milovskaya et al. /129a/ studied the reaction between $Al(C_2H_5)_3$ and benzoyl peroxide at lower concentrations than Razuvaev et al. /129/; the identity of the final reaction products indicated the formation of diethylaluminum benzoate and free ethyl radicals.

Trialkylaluminum compounds react with sulfur, selenium and tellurium, under relatively mild conditions with formation of corresponding derivatives /130/:

$$AlR_3 + nX \longrightarrow R_{3-n}Al(XR)_n$$

where X = S, Se, Te; $n = 1, 2, 3$ and R is an alkyl. Thus, the following compounds have been prepared:

$$Al(C_2H_5)_3 + S \xrightarrow[70-110°C]{hexane} (C_2H_5)_2AlSC_2H_5$$

$$2Al(C_2H_5)_3 + 2Se \xrightarrow{160°C} (C_2H_5)_2AlSeC_2H_5 + C_2H_5AlSe(C_2H_5)_2$$

The following alkoxy and alkylthio derivatives are formed with alcohols and mercaptans /17/:

$$Al(CH_3)_3 + 3CH_3OH \longrightarrow Al(OCH_3)_3 + 3CH_4$$

$$Al(CH_3)_3 + CH_3SH \longrightarrow (CH_3)_2AlSCH_3 + CH_4$$

Alkylaluminum compounds react at moderate temperatures with ethers and thioethers to form complex compounds such as $AlR_3 \cdot OR'_2$ or $AlR_3 \cdot SR'_2$.

Alkylaluminum compounds may react with CO_2 in two ways. At room temperature and in the presence of excess trialkylaluminum one Al — C bond reacts:

$$(C_2H_5)_3Al + CO_2 \longrightarrow (C_2H_5)_2Al{-}O{-}COC_2H_5$$

$$(C_2H_5)_2Al{-}O{-}COC_2H_5 + 2(C_2H_5)_3Al \longrightarrow$$

$$\longrightarrow \underset{\text{alumoxan}}{(C_2H_5)_2AlOAl(C_2H_5)_2} + (C_2H_5)_3COAl(C_2H_5)_2$$

$$(C_2H_5)_3COAl(C_2H_5)_2 \xrightarrow{H_2O} \underset{\text{triethylcarbinol}}{(C_2H_5)_3COH}$$

At a high temperature and in the presence of excess CO_2 derivatives of carboxylic acids are formed:

$$(C_2H_5)_3Al + 2CO_2 \longrightarrow C_2H_5Al(OCOC_2H_5)_2$$

$$C_2H_5Al(OCOC_2H_5)_2 \xrightarrow{H_2O} 2C_2H_5COOH$$

The last mentioned reaction may be utilized in the preparation of various straight-chain acids with odd and even numbers of carbon atoms:

$$(C_2H_5)_3Al + 3nC_2H_4 \longrightarrow [H(CH_2CH_2)_{n+1}]_3Al \xrightarrow{CO_2+H_2O} C_2H_5(CH_2CH_2)_nCOOH$$

$$(C_3H_7)_3Al + 3nC_2H_4 \longrightarrow [CH_3(CH_2CH_2)_{n+1}]_3Al \xrightarrow{CO_2+H_2O} CH_3(CH_2CH_2)_{n+1}COOH$$

Dialkylaluminum halides and alkylaluminum dihalides do not react with CO_2 /22/. Triethylaluminum reacts with SO_2 at −75°C to form the aluminum salt of ethylsulfinic acid:

$$Al(C_2H_5)_3 + 3SO_2 \longrightarrow Al(O_2SC_2H_5)_3$$

Many reactions between alkylaluminum compounds and oxygenated organic compounds have been realized. The composition of the reaction products depends on the initial ratio between the reducing and the oxidizing agents. The following reactions can be realized in high yields /101, 131 — 135/:

Ketones → Secondary alcohols
Aldehydes → Primary alcohols
Ethers → Secondary alcohols

Esters	→ Aldehydes or primary alcohols
Benzoic acid	→ Benzyl alcohol
Ortho-esters	→ Acetals
Acetals and ketals	→ Ethers

Dialkylaluminum hydride reacts with alkenyl ethers with formation of olefins /136/:

$$R{-}O{-}CH{=}CHR' + HAlR''_2 \longrightarrow ROAlR''_2 + CH_2{=}CHR'$$

The reaction between tribenzylaluminum and formaldehyde followed by the hydrolysis of the reaction product yields phenylethanol $C_6H_5CH_2CH_2OH$ /137/. Alkylaluminum dichlorides react with acid chlorides and carbalkoxy acid chlorides to yield the corresponding ketones and keto-esters /138/:

$$RAlCl_2 + R'COCl \longrightarrow R'COR + AlCl_3$$

where $R = CH_3$, C_2H_5; $R' = C_6H_5$, C_3H_7, C_4H_9.

$$R'AlCl_2 + R''COCl \longrightarrow R''COR' + AlCl_3$$

where $R' = CH_3$, C_2H_5; $R'' = C_2H_5COO(CH_2)_8$, $C_2H_5COO(CH_2)_4$, $C_2H_5COO(CH_2)_2$.

The reduction of Si — O bonds by dialkylaluminum hydride yields silanes and the corresponding alcoholates /139/:

$$R_2AlH + R'_3SiOCH{=}CH_2 \longrightarrow R_2AlOCH{=}CH_2 + R'_3SiH$$

Trialkylaluminum compounds react with hexaalkyldisiloxane in a similar manner:

$$Al(C_nH_{2n+1})_3 + R_3SiOSiR_3 \longrightarrow R_3SiH + (C_nH_{2n+1})_2AlOSiR_3 + C_nH_{2n}$$

Silanes are formed in high yields. The reaction between dimethylaluminum bromide and disiloxane yields thermally stable, dimeric, volatile tetramethyldisiloxyaluminum /140/:

$$Al_2(CH_3)_4Br_2 + 2(SiH_3)_2O \longrightarrow (CH_3)_4Al_2(OSiH_3)_2 + 2SiH_3Br$$

Polysiloxanes may be reacted with alkylaluminum halides to form tetraalkylsilanes /141/.

Chromium halides react with triethylaluminum and carbon monoxide to form chromium hexacarbonyl /142/.

Reactions with elements of Group VII and their compounds

The action of halogens results in a vigorous dealkylation of alkylaluminum compounds /143/:

$$AlR_3 + Cl_2 \longrightarrow R_2AlCl + RCl$$

Reactions with other halogens take place in a similar manner.

Reactions between alkyl halides and alkylaluminum compounds may proceed in various manners. At a low temperature isopropyl bromide and butyl bromide cause the ethyl groups of triethylaluminum to be substituted by bromine /144/:

$$Al(C_2H_5)_3 + C_4H_9Br \longrightarrow BrAl(C_2H_5)_2 + C_6H_{14}$$

Dolgoplosk et al. /4/ failed to note a reaction of this kind between triethylaluminum and ethyl chloride.

Reactions between alkylaluminum compounds and polyhalo derivatives proceed very vigorously, with possibility of explosions /159/.

Reactions with compounds containing mobile hydrogen

All organoaluminum compounds vigorously react with acids, water, alcohol, thioalcohols, phenols and similar compounds with formation of the corresponding hydroxy derivatives of aluminum and hydrocarbons /1, 26, 86/. When these reagents are in excess, three hydrocarbon groups are split off:

$$AlR_3 + 3R'COOH \longrightarrow (R'COO)_3Al + 3RH$$

$$AlR_3 + 3H_2O \longrightarrow Al(OH)_3 + 3RH$$

$$AlR_3 + 3R'OH \longrightarrow Al(OR')_3 + 3RH$$

$$AlR_3 + 3R'SH \longrightarrow Al(SR')_3 + 3RH$$

$$AlR_3 + 3ArOH \longrightarrow Al(OAr)_3 + 3RH$$

If the reagent ratio is smaller, one alkyl group only may be substituted, for example /145, 146/:

$$CH_3COOH + Al(\text{iso-}C_4H_9)_3 \longrightarrow CH_3COOAl(\text{iso-}C_4H_9)_2 + \text{iso-}C_4H_{10}$$

If water is deficient, the hydrolysis results in the formation of alkylalumoxanes /147, 148/:

$$2AlR_3 + H_2O \longrightarrow R_2Al{-}O{-}AlR_2 + 2RH$$

The reaction very probably involves the formation of R_2Al-OH as an intermediate product. Reaction between tertiary alcohols and trialkylaluminum compounds results in the splitting off of one alkyl group only /149/:

$$AlR_3 + R'_3COH \longrightarrow R_2AlOCR'_3 + RH$$

Dialkylaluminum hydrides react with alkali metal hydroxides with the formation of dialkylaluminumoxymetals:

$$R_2AlH + MOH \longrightarrow R_2AlOM + H_2$$

These compounds, which are readily soluble in organic solvents, may be utilized as aluminum activators in the direct synthesis of trialkylaluminum compounds.

3. SYNTHESIS OF ORGANOALUMINUM COMPOUNDS

Various methods for the preparation of organoaluminum compounds and their derivatives have been described in the literature. In what follows we shall give the methods for the preparation of alkylaluminum compounds, under eight different headings, and shall give a brief account of their technological properties.

Reaction between alkyl (aryl) halides and metallic aluminum

The reaction $2Al + 3RX \rightarrow AlR_2X + AlRX_2$, which was first realized in 1859 by Hollwachs and Schafarik /150/, yields various alkyl(aryl)aluminum sesquihalides, and in certain cases dialkylaluminum halides and trialkylaluminum compounds.

Methyl- and ethylaluminum sesquihalides are now produced in this way on an industrial scale. Aluminum is used as powder or as shavings, or else as a mixture or alloy with magnesium. The alkyl halides employed include chlorides, bromides and iodides. Aluminum may be activated by iodine, bromine, alkylaluminum halides, mercury halides, titanium halides, aluminum halides, alkyl iodides and alkyl bromides, as well as aluminum alloyed with lithium, copper, calcium and zinc.

The synthesis is carried out in a saturated hydrocarbon medium, in ethers or in alkylaluminum sesquihalides. It was suggested by a number of workers that the synthesis of alkylaluminum halides be performed in the gas phase, by reacting alkyl halide vapor with aluminum in the presence of primers /151, 152/. Better results are obtained in the preparation of lower alkylaluminum compounds — trimethylaluminum, triethylaluminum, methylaluminum sesquihalides and ethylaluminum sesquihalides. The preparation of higher alkylaluminum compounds by this method is accompanied by side reactions, which may be suppressed by conducting the reaction in anisole /22/.

Dehalogenation of alkylaluminum halides

The dehalogenation of alkylaluminum halides is usually effected with the aid of alkali and alkaline earth metals, and by alkali metal halides. The reactions can be generally written as follows:

$$AlR_3 \cdot AlX_3 + 3Na \longrightarrow AlR_3 + 3NaX + Al$$

$$AlR_3 \cdot AlCl_3 + NaCl \longrightarrow AlR_2Cl + NaCl \cdot AlRCl_2$$

In this way trimethylaluminum, triethylaluminum, diethylaluminum chloride, tripropylaluminum, tributylaluminum, etc., can all be prepared. The dehalogenation agents include metallic sodium, sodium amalgam, potassium-sodium alloys, potassium and sodium chlorides, magnesium-aluminum alloys, metallic magnesium and sodium fluoride. The alkyl-aluminum compounds obtained as a result of the reaction contain admixtures of alkylaluminum halides, which must be removed by conducting a supplementary dehalogenation. Major amounts of metal halides are obtained as side products /1/. Dehalogenation by metallic sodium is used in small-scale industrial production of trimethyl- and triethylaluminum and of diethylaluminum chloride.

Reactions between organic compounds of Group I — IV elements and aluminum halides or metallic aluminum

Laboratory syntheses of various organoaluminum compounds based on reactions between metallic aluminum, aluminum halides and alkylaluminum halides and organic compounds of elements in Groups I — IV have been described /1, 86/.

Thus, for instance, lithium aluminum tetraalkyls react with aluminum chloride to form trialkylaluminum compounds /153/:

$$3LiAlR_4 + AlCl_3 \longrightarrow 4AlR_3 + 3LiCl$$

In the presence of polar solvents the following syntheses are possible /154/:

$$AlCl_3 + RM \longrightarrow RAlCl_2 + MCl$$
$$AlCl_3 + 2RM \longrightarrow R_2AlCl + 2MCl$$
$$AlCl_3 + 3RM \longrightarrow AlR_3 + 3MCl$$
$$(M = Li,\ Na)$$

The use of organic derivatives of lithium and sodium gave the etherates $(n\text{-}C_4H_9)_2AlCl$, $Al(C_6H_5)_3$ and $C_2H_5AlCl_2$ in low yields. Organoaluminum compounds could not be obtained by the reaction between $n\text{-}C_4H_9Na$ and $AlCl_3$ in mineral oil. Syntheses of trialkynylaluminum compounds have been performed in tetrahydrofuran, dioxan, trimethylamine and pyridine, according to the reaction /106, 155/:

$$AlCl_3 \cdot B + 3Na{-}C{\equiv}C{-}R \longrightarrow Al(C{\equiv}CR)_3 \cdot B + 3NaCl$$

where B is ethyl ether, dioxan, tetrahydrofuran, trimethylamine or pyridine, and $R = CH_3$, iso-C_4H_9, C_6H_5.

The reaction

$$2Al + 3HgR_2 \longrightarrow 2AlR_3 + 3Hg$$

in ether served to prepare etherates of trimethylaluminum, triethyl-aluminum and tripropylaluminum in high yields; the duration of the reaction was 24 hours /9, 156/. When this reaction was effected in xylene for

40 hours, the following compounds were obtained in yields of up to 50%: triisobutylaluminum, tridecylaluminum, triphenylaluminum and tritolylaluminum /157/.

Sodium aluminum tetraethyl reacts with ethylaluminum sesquichloride with formation of triethylaluminum /158/:

$$3NaAl(C_2H_5)_4 + Al_2(C_2H_5)_3Cl_3 \longrightarrow 5Al(C_2H_5)_3 + 3NaCl$$

It was recently reported by Ziegler /67/ that aluminum in the presence of mercury reacts with sodium aluminum tetraalkyls and with sodium aluminum alkoxytrialkyls with formation of trialkylaluminum compounds and dialkylaluminum oxides:

$$3NaAlR_4 + Al + 3Hg \longrightarrow 4AlR_3 + 3HgNa$$
$$3NaAlR_3(OR) + Al + 3Hg \longrightarrow 3AlR_2(OR) + AlR_3 + 3HgNa$$

These reactions can be realized in various solvents at 85 — 160°C.

It was established by Zakharkin and Okhlobystin /28/ that reactions between metallic aluminum and diisoalkylmercury yield mixtures of organoaluminum compounds with normal and branched chain hydrocarbon radicals. On prolonged heating a full rearrangement takes place, and all radicals in the product assume normal structure.

A complex compound of triperfluorovinylaluminum with trimethylamine was prepared in ether as follows /160/:

$$3(CF_2{=}CF)_2Hg + 2AlH_3 \cdot N(CH_3)_3 \xrightarrow{-20\ ^\circ C} 2(CF_2{=}CF)_3Al \cdot N(CH_3)_3 + 3Hg + 3H_2$$

In tetrahydrofuran, dioxan, trimethylamine and pyridine the following compounds can be prepared /155/:

$$2Al + 3Hg(C{\equiv}CR)_2 \longrightarrow 2Al(C{\equiv}CR)_3 + 3Hg$$
$$2AlH_3 \cdot N(CH_3)_3 + 3Hg(C{\equiv}CR)_2 \longrightarrow 2Al(C{\equiv}CR)_3 \cdot N(CH_3)_3 + 3Hg + 3H_2$$

The latter reaction involves partial hydrogenation of the triple bond:

$$al{-}C{\equiv}CR + alH \longrightarrow \begin{matrix} al \\ al \end{matrix}\!\!>\!C{=}CHR^{*}$$

The aluminum derivatives of cyclopentadiene, methylcyclopentadiene and indene have been synthesized in toluene in high yields at 45°C, using aluminum activated by aluminum chloride /161, 162/:

$$2Al + 3(C_5H_5)_2Hg \longrightarrow \underset{\text{m.p. } 50-60\ ^\circ C}{2Al(C_5H_5)_3} + 3Hg$$
$$2Al + 3(CH_3C_5H_4)_2Hg \longrightarrow 2Al(C_5H_4CH_3)_3 + 3Hg$$
$$2Al + 3(C_9H_7)_2Hg \longrightarrow 2Al(C_9H_7)_3 + 3Hg$$

Organomagnesium compounds gave high yields of triethylaluminum etherates /163/:

* Here and in what follows al = 1/3 Al.

$$(C_2H_5)_2AlBr + C_2H_5MgCl \xrightarrow{(C_2H_5)_2O} Al(C_2H_5)_3 \cdot (C_2H_5)_2O + MgClBr$$

The optically active etherate of tris(δ-2-methylbutyl)-aluminum has been prepared /164, 165/:

$$H_3C{-}\underset{C_2H_5}{\overset{CH_2OH}{C^*H}} \xrightarrow{SOCl_2} H_3C{-}\underset{C_2H_5}{\overset{CH_2Cl}{C^*H}} \xrightarrow{Mg} H_3C{-}\underset{C_2H_5}{\overset{CH_2MgCl}{C^*H}} \xrightarrow{AlCl_3}$$

$$\longrightarrow Al\left(CH_2{-}\underset{CH_3}{\overset{H}{C^*}}{-}C_2H_5\right)_3$$

Many vinyl compounds of aluminum have recently been prepared, including $(CH_2=CH)_2AlCl$, $(CH_2=CH)_2AlBr$, $(CH_2=CH)_2AlOCH_3$, $(CH_2=CH)_2AlCH_3$, $(CH_2=CH)_3Al$, $(CH_2=CH)_2AlC_6H_5$, etc. /166/.

Organozinc compounds have been used in the preparation of diethylaluminum chloride and ethylaluminum sesquichloride /167, 168/:

$$AlCl_3 + 2C_2H_5ZnCl \longrightarrow (C_2H_5)_2AlCl + 2ZnCl_2$$

The yield was 91% on $AlCl_3$.

Triarylaluminum compounds are readily obtained by the reaction /26/:

$$BAr_3 + AlR_3 \longrightarrow AlAr_3 + BR_3$$

Phenylaluminum dichloride is formed in satisfactory yields by the reaction between aluminum chloride and phenylchlorosilanes /169/:

$$AlCl_3 + C_6H_5SiCl_3 \longrightarrow C_6H_5AlCl_2 + SiCl_4$$
$$AlCl_3 + (C_6H_5)_2SiCl_2 \longrightarrow C_6H_5AlCl_2 + C_6H_5SiCl_3$$
$$2AlCl_3 + (C_6H_5)_2SiCl_2 \longrightarrow 2C_6H_5AlCl_2 + SiCl_4$$

Organoaluminum compounds containing five-membered and six-membered hetero rings have been prepared as follows /170/:

$$(\text{iso-}C_4H_9)_2AlOCH_2CH_2CH{=}CH_2 \xrightarrow{HAl(\text{iso-}C_4H_9)_2}$$

$$\longrightarrow \text{iso-}C_4H_9Al\langle{}^{O-CH_2}_{CH_2-CH_2}\rangle CH_2 + \text{iso-}C_4H_8$$

$$(\text{iso-}C_4H_9)_2Al\underset{C_2H_5}{N}CH_2CH{=}CH_2 \xrightarrow{HAl(\text{iso-}C_4H_9)_2} \text{iso-}C_4H_9Al\langle{}^{CH_2-CH_2}_{\underset{C_2H_5}{N}-CH_2}\rangle + \text{iso-}C_4H_8$$

The inner complexes

$$(iso\text{-}C_4H_9)_2Al \overset{CH_2-CH_2}{\underset{X-CH_2}{\leftarrow}} CH_2 \quad \text{and} \quad (iso\text{-}C_4H_9)_2Al \overset{CH_2-CH_2}{\underset{X-CH_2}{\leftarrow}}$$

where X is OC_2H_5 or $N(C_2H_5)_2$, are formed by the reaction between triethylaluminum or diisobutylaluminum hydride and the corresponding butenyl derivatives $CH_2 = CH(CH_2)_nX$ ($n = 2, 3$) /171/.

The preparation of the peroxy compound diethoxyaluminumperoxycumyl has been reported /127, 128/. Alkylaluminum compounds with hetero atoms in alkyl hydrocarbon chains have also been prepared /172/:

$$(CH_3)_3SiCH_2CH{=}CH_2 + (iso\text{-}C_4H_9)_2AlH \longrightarrow (iso\text{-}C_4H_9)_2AlCH_2CH_2CH_2Si(CH_3)_3$$

$$3(iso\text{-}C_4H_9)_2AlCH_2CH_2CH_2Si(CH_3)_3 \longrightarrow Al[CH_2CH_2CH_2Si(CH_3)_3]_3 + 2Al(iso\text{-}C_4H_9)_3$$

Reactions of trialkylaluminum compounds and aluminum with aluminum halides, aluminum alcoholates and ethers

Alkyl(aryl)aluminum halides of various degrees of alkylation may be prepared by the reaction between alkylaluminum or arylaluminum compounds and aluminum halides /69, 73/:

$$2AlR_3 + AlX_3 \longrightarrow 3AlR_2X$$

$$AlR_3 + 2AlX_3 \longrightarrow 3AlRX_2$$

$$Al_2R_3X_3 + AlX_3 \longrightarrow 3AlRX_2$$

Their apparent simplicity notwithstanding, the reactions are accompanied by side processes, with the result that the main products are not obtained pure.

Alkylaluminum fluorides are also obtained by the reaction between alkylaluminum chlorides with fluorides of alkali or alkaline earth metals /70, 71/:

$$Al(C_2H_5)_2Cl + NaF \longrightarrow Na[Al(C_2H_5)_2Cl]F \longrightarrow Al(C_2H_5)_2F + NaCl$$

$$2Al(C_2H_5)_2Cl + CaF_2 \longrightarrow 2Al(C_2H_5)_2F + CaCl_2$$

$$2Al(C_2H_5)_2Cl + BaF_2 \longrightarrow 2Al(C_2H_5)_2F + BaCl_2$$

$$2Al_2(C_2H_5)_3Cl_3 + 3CaF_2 \longrightarrow 2Al_2(C_2H_5)_3F_3 + 3CaCl_2$$

These reactions take place at high temperatures — 180 to 200°C — and the yields of the final products may attain up to 85%. Moreover, the synthesis of alkylaluminum halides may be realized with metallic aluminum and halogens in accordance with the following equations /173, 174/:

$$4AlR_3 + 2Al + 3X_2 \longrightarrow 6R_2AlX$$

$$2AlR_3 + 2Al + 3X_2 \longrightarrow 2Al_2R_3X_3$$

$$AlR_3 + 2Al + 3X_2 \longrightarrow 3AlRX_2$$

Patent literature gives examples of the syntheses of $Al(C_2H_5)_2I$, $Al(C_2H_5)_2Br$, $Al_2(C_2H_5)_3Br_3$, $Al(CH_3)Br_2$, $Al(C_3H_7)_2I$ in 87.5 — 93% yields. Methylenealuminum bromide reacts with iodine according to the following equation /175/:

$$BrAl{=}CH_2 + I_2 \longrightarrow IBrAlCH_2I$$

Methylaluminum dichloride has been obtained in 76% yield /176/ by the reaction between metallic aluminum and methyl ether and aluminum chloride. Methoxymethylaluminum chloride is probably formed first and disproportionates when distilled in vacuo, with the separation of methylaluminum dichloride:

$$AlCl_3 + 2Al + 3CH_3OCH_3 \longrightarrow 3CH_3Al(OCH_3)Cl$$
$$3CH_3Al(OCH_3)Cl \longrightarrow Al(CH_3)Cl_2 + Al_2(CH_3)_2(OCH_3)_3Cl$$

Alkoxyalkylaluminum halides AlR(OR)X have been obtained by reactions between alkylaluminum sesquihalides and aluminum alcoholates /177/ or between metallic aluminum and aluminum chloride and ethers /176, 178/.

Reactions of aluminum hydrides, alkali metal hydrides and aluminum halides with olefins

One of the fundamental reactions of this type is $AlH_3 + 3CH_2 = CR'R \rightarrow Al(CH_2CHRR')_3$. It takes place easily and rapidly with many olefins /26, 179, 180/. However, its practical possibilities are limited, since aluminum hydride is difficult to prepare and its thermal stability is low.

The reaction between dialkylaluminum hydrides and olefins is employed to prepare mixed trialkylaluminum compounds /153, 179/:

$$(C_2H_5)_2AlH + CH_3CH{=}CH{-}CH_2{-}CH_3 \xrightarrow{70°C} (C_2H_5)_2Al{-}CH(CH_3)CH_2CH_2CH_3$$

$$(C_2H_5)_2AlH + \text{[bicyclic olefin with } CH_2 \text{ bridges]} \xrightarrow[\text{hexane}]{65°C} (C_2H_5)_2Al{-}\text{[bicyclic with } CH_2 \text{ bridges]}$$

$$(C_2H_5)_2AlH + CH_2{=}CH{-}C_6H_5 \xrightarrow{65°C} (C_2H_5)_2AlCH_2CH_2C_6H_5$$

$$2(C_2H_5)_2AlH + CH_2{=}CH{-}CH{=}CH_2 \longrightarrow (C_2H_5)_2AlCH_2CH_2CH_2CH_2Al(C_2H_5)_2$$

$$(C_2H_5)_2AlH + \text{cyclopentene} \longrightarrow (C_2H_5)_2Al{-}\text{cyclopentyl}$$

$$(C_2H_5)_2AlH + \text{cyclohexadiene} \longrightarrow (C_2H_5)_2Al{-}\text{cyclohexenyl}$$

The composition of the end products depends on the structure of the alkylaluminum compound and of the olefin. Thus, for instance, dialkylaluminum hydride and styrene may react in two ways:

$$R_2AlH + CH_2{=}CHC_6H_5 \longrightarrow R_2AlCH_2CH_2C_6H_5$$
$$R_2AlH + \underset{\displaystyle \| \atop CH_2}{CH}{-}C_6H_5 \longrightarrow R_2Al{-}\underset{\displaystyle | \atop CH_3}{CH}{-}C_6H_5$$

Various trialkylaluminum compounds and their derivatives may be prepared by the interaction of aluminum halides with olefins and metal hydrides /44, 67, 181—184/. The general equation for these reactions is:

$$3MH + AlX_3 + 3C_nH_{2n} \longrightarrow Al(C_nH_{2n+1})_3 + 3MX$$

When the ratio between the metal hydride, aluminum halide and olefin is changed, various products are obtained, e. g.:

$$AlX_3 + 3MH + 2C_2H_4 \longrightarrow HAl(C_2H_5)_2 + 3MX$$
$$AlX_3 + 2MH + 2C_2H_4 \longrightarrow Al(C_2H_5)_2X + 2MX$$
$$AlX_3 + 4MH + C_2H_4 \longrightarrow MAl(C_2H_5)X_3$$

These reactions take place in ether and in hydrocarbon solvents at high yields and low temperatures and may be used to prepare $Al(C_2H_5)_2H$, $Al(C_3H_7)_2H$, $Al(C_6H_{13})_2H$, and other compounds /44, 153, 182/.

Electrolysis of complex organoaluminum compounds

Recent patent literature /107/ contains a description of the synthesis of triethylaluminum, which proceeds as follows.

1. Preparation of metal aluminum tetraalkyl:

$$3MH + 3Al(C_2H_5)_3 + 3C_2H_4 \longrightarrow 3MAl(C_2H_5)_4$$

2. Electrolysis of metal aluminum tetraalkyl:

$$3MAl(C_2H_5)_4 + Al + 3Hg \longrightarrow 4Al(C_2H_5)_3 + 3MHg$$

3. Electrolysis of metal amalgam:

$$3MHg \longrightarrow 3M + 3Hg$$

4. Preparation of metal hydride:

$$3M + 1^1/_2H_2 \longrightarrow 3MH$$

The overall electrochemical process corresponds to the direct synthesis of triethylaluminum:

$$Al + 1^1/_2H_2 + 3C_2H_4 \longrightarrow Al(C_2H_5)_3$$

The following electrolytes have been proposed as substitutes for the metal aluminum tetraethyl: $NaAl(C_2H_5)_3F$, $NaAl(C_2H_5)_4$, $NaAl(C_2H_5)_3OC_4H_9$, mixture of $NaF \cdot 2Al(C_2H_5)_3$ with $NaAl(C_2H_5)_4$, etc. In order to prevent thermal decomposition of the trialkylaluminum compound formed, it is recommended that the electrolysis be performed in vacuo or with the use of extractants which are immiscible with the electrolyte.

Electrochemical preparation of triethylaluminum is schematically represented as a cyclic process, consisting of five complex technological stages:

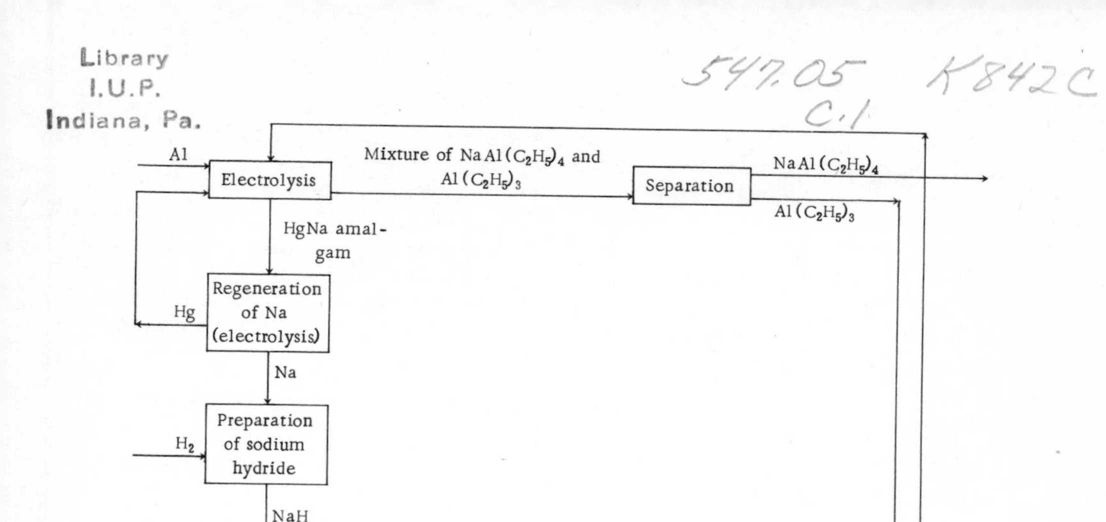

Ziegler proposed similar reaction schemes for the electrolytic preparation of organic compounds of zinc, tin, cadmium, etc.

Direct synthesis from metallic aluminum, hydrogen and olefins

In a brief communication /185/, Ziegler reported in 1955 the realization of the direct synthesis of trialkylaluminum compounds. This was the beginning of numerous investigations in the field of organoaluminum compounds; a new interesting field in the chemistry of organometallic compounds was discovered and gave rise to new branches of petrochemical industry. The synthesis and the chemistry of these compounds now form the subject of hundreds of patents and publications.

Various trialkylaluminum compounds and dialkylaluminum hydrides can be prepared by rapid technological synthesis from cheap and readily available raw materials — aluminum, hydrogen and olefins — by direct synthesis in a wide range of temperatures (30 to 300°C) and pressures (5 to 300 atm).

The reaction of direct synthesis

$$Al + 1^1/_2H_2 + 3CH_2{=}CHR \longrightarrow Al(CH_2CH_2R)_3$$

takes place only in the presence of the trialkylaluminum in the reactant mixture, which means that the compound participates as the intermediate product in the individual stages of the direct synthesis. In a recent synthesis, isoprene was used as the olefin in the direct synthesis of isoprenylaluminum /186/:

$$Al + 3C_5H_8 + 1^1/_2H_2 \longrightarrow Al(C_5H_9)_3$$

Isoprenylaluminum in conjunction with transition metal salts proved to be an active catalyst for polymerization of ethylene /187, 188/.

Certain olefins, such as ethylene, which are capable of participating in the side reaction of chain extension in the alkyl radicals under the experimental conditions of the direct synthesis, form an important exception. In this case, especially so in the preparation of triethylaluminum, a two-stage process takes place. The first stage consists in the interaction between aluminum, hydrogen and trialkylaluminum with formation of dialkylaluminum hydride:

$$Al + 1^1/_2 H_2 + 2Al(CH_2CH_2R)_3 \longrightarrow 3HAl(CH_2CH_2R)_2$$

In the second stage the dialkylaluminum hydride is alkylated by the olefin with formation of trialkylaluminum:

$$3HAl(CH_2CH_2R)_2 + 3CH_2{=}CHR \longrightarrow 3Al(CH_2CH_2R)_3$$

Intensive attempts have been made abroad to develop a one-stage synthesis of triethylaluminum, but the results, which appeared in a number of patents, are contradictory.

The only alkylaluminum halides which have so far been directly prepared are alkylaluminum sesquihalides. In 1937 Hall and Nash /189/ obtained a product similar to ethylaluminum sesquichloride while attempting to polymerize ethylene in the presence of aluminum chloride and metallic aluminum. Ruthruff /190/ in 1942 recommended that alkylaluminum sesquihalides be directly synthesized according to the following scheme:

$$Al + AlX_3 + 1^1/_2 H_2 + 3C_nH_{2n} \longrightarrow Al_2(C_nH_{2n+1})_3X_3$$

Many workers reported that aluminum must be activated prior to being used in direct synthesis. According to Ziegler, metallic aluminum does not react by itself unless the passivating oxide film on its surface has been previously removed in some way /26/. Aluminum may be activated in different ways, e.g., by grinding in ball mills and vibratory mills in an inert atmosphere, sometimes in the presence of activating agents such as aluminum chloride, halogens, higher organic acids and other compounds.

Details of direct synthesis of alkylaluminum compounds and modes of activation of the aluminum used for this purpose will be found in the following sections.

Reactions of organoaluminum compounds with olefins and acetylenes

New organoaluminum compounds can be prepared by the following three varieties of these reactions:

1) displacement:

$$Al(C_nH_{2n+1})_3 + 3C_mH_{2m} \longrightarrow Al(C_mH_{2m+1})_3 + 3C_nH_{2n}$$

2) chain extension:

$$AlR_3+3nC_2H_4 \longrightarrow Al[(C_2H_4)_nR]_3$$

3) addition:

$$AlR_2Cl+NaC{\equiv}C{-}R \longrightarrow AlR_2C{\equiv}C{-}R+NaCl$$

The initial material in displacement reactions is usually the readily available triisobutylaluminum. When this compound is made to react with an olefin at an elevated temperature, various alkylaluminum compounds are formed in high yields. In certain cases the reactions are conducted under elevated pressures and in the presence of nickel acetylacetonate, alkylaluminum compounds or alumosilicates as catalysts.

bis-(Diisobutylaluminum) isopentene has been recently obtained as follows /191/:

$$2(\text{iso-}C_4H_9)_3Al+CH_2{=}C(CH_3)CH{=}CH_2 \xrightarrow{78-88°C}$$
$$\longrightarrow (\text{iso-}C_4H_9)_2AlCH_2CH(CH_3)CH_2CH_2Al(\text{iso-}C_4H_9)_2+2\,\text{iso-}C_4H_8$$

Such compounds have a high hydrolytic and thermal stability and react with transition metal salts to form olefin polymerization catalysts.

Numerous studies of the chain extension reaction (see, for instance, /26, 88 — 95, 97, 192, 194/) mainly concern the reactions of ethylene with trialkylaluminum compounds, as a result of which mixtures of higher alkylaluminum compounds are formed.

Ziegler and Gellert /179/ also reported the addition of octene-1 to triethylaluminum, resulting in the formation of the triisodecylaluminum $Al[CH_2-CH(C_2H_5)(CH_2)_5CH_3]_3$. Reactions between alkylaluminum compounds and acetylenes and diolefins may yield alkenylaluminum compounds and bifunctional compounds. Nesmeyanov et al. /195/ studied the reaction of tolane with triethylaluminum and found that the following products are formed in the course of the reaction:

$$\underset{(I)}{C_6H_5{-}C{\equiv}C{-}C_6H_5}+Al(C_2H_5)_3 \longrightarrow \underset{(II)}{C_2H_5{-}C(C_6H_5){=}C(C_6H_5){-}Al(C_2H_5)_2}$$

$$(I)+(II) \longrightarrow (C_6H_5)(C_2H_5)C{=}C(C_6H_5){-}C(C_6H_5){=}C(C_6H_5){-}Al(C_2H_5)_2$$

The addition reaction is also of interest in connection with the reactions between alkylaluminum compounds and dienic and acetylenic hydrocarbons. Thus, butadiene reacts with diethylaluminum hydride to form butenyl-2-diethylaluminum and bis(diethylaluminum)butylene /179/:

$$(C_2H_5)_2AlH+CH_2{=}CH{-}CH{=}CH_2 \longrightarrow (C_2H_5)_2AlCH_2{-}CH{=}CH{-}CH_3$$
$$2(C_2H_5)_2AlH+CH_2{=}CH{-}CH{=}CH_2 \longrightarrow (C_2H_5)_2AlCH_2CH_2CH_2CH_2Al(C_2H_5)_2$$

Alkyl-substituted hexadienes, such as 2, 5-dimethylhexadiene-1, 5, react with triisobutylaluminum with the formation of tricycloalkylaluminum /180/:

$$3\left|\begin{matrix}CH_2-\overset{\overset{CH_3}{|}}{C}=CH_2\\CH_2-\underset{\underset{CH_3}{|}}{C}=CH_2\end{matrix}\right. + Al(iso\text{-}C_4H_9)_3 \xrightarrow{3\ iso\text{-}C_4H_8} \left[\begin{matrix}CH_2-\overset{\overset{CH_3}{|}}{C}=CH_2\\|\\CH_2-\underset{\underset{CH_3}{|}}{CH}-CH_2-\end{matrix}\right]_3 Al \longrightarrow$$

$$\longrightarrow \left[\begin{matrix}CH_2-\overset{\overset{CH_3}{|}}{C}-CH_2-\\| \qquad\quad |\\CH_2-\underset{\underset{CH_3}{|}}{CH}-CH_2\end{matrix}\right]_3 Al$$

This field of organoaluminum chemistry has not yet been adequately studied; in our view, the possibilities of utilization of this class of compounds in organic chemistry in this manner are practically unlimited.

Since some of the above-mentioned methods of preparation of alkylaluminum compounds are now extensively used in industry, the following sections will deal with the details of preparation of these compounds by direct synthesis from aluminum, hydrogen and olefins, and also with the reactions of aluminum, alkyl halides and certain alkali and alkaline earth metals.

4. SYNTHESIS OF ALKYLALUMINUM COMPOUNDS FROM ALUMINUM, HYDROGEN AND OLEFINS

Hall and Nash /189/ were the first to report the possibility of direct synthesis of organoaluminum compounds. These workers reacted metallic aluminum, ethylene and aluminum chloride in petroleum ether at about 150°C under 50 — 90 atm pressure and obtained ethylaluminum sesquichloride as a result. The formation of a mixture of organoaluminum compounds most probably takes place in this case by several successive reactions, as postulated by Nesmeyanov and Sokolik /86/. The first stage is the formation of HCl by the reaction between aluminum chloride and the hydrocarbon. Hydrogen chloride reacts with ethylene to form ethyl chloride, which in turn reacts with aluminum to give ethylaluminum sesquichloride. It should be noted that the above workers failed to note the formation of higher organoaluminum compounds during the synthesis; according to Ziegler and Kroll /26/, this is due to the weak reactivity of alkylaluminum halides toward ethylene. If ethylene is used as the olefin in the direct synthesis of alkylaluminum compounds from hydrogen, aluminum and α-olefin in the presence of a trialkylaluminum compound, the reaction

$$Al + 1^1/_2H_2 + 3C_2H_4 \longrightarrow Al(C_2H_5)_3$$

will not be limited to the formation of triethylaluminum, and at higher temperatures higher trialkylaluminum compounds will be formed /185/:

$$alC_2H_5 + nC_2H_4 \longrightarrow al(C_2H_4)_nC_2H_5$$

A few patents were subsequently taken out for the synthesis of triethylaluminum by the interaction between aluminum, hydrogen and ethylene. Thus, American and British patents /196, 197/ have been taken out for one-step synthesis of triethylaluminum; the yield is quantitative and the working conditions are 30 — 130°C, total pressure 15 — 300 atm, and partial pressure of ethylene 10 — 100 atm. The synthesis was complete after 0.5 — 20 hours, while the reaction rate attained 2 moles of $Al(C_2H_5)_3$ per 1 gram-atom Al per hour. A Japanese patent /198, 199/ gives the results of a study of one-step synthesis of triethylaluminum, conducted at different values of overall pressure, ethylene: hydrogen ratio, temperature and duration of process. The experimental data reported in these patents indicate that triethylaluminum is formed under an overall pressure of 50 — 250 atm, and only within the following limiting values of temperature and partial ethylene pressure:

Reaction temperature, °C	80	130	200
Partial pressure of ethylene, atm	10	167	167

It was shown in subsequent foreign publications that other α-olefins react with alkylaluminum compounds in different manners, both in the presence and in the absence of aluminum. It was noted, for example, that all α-olefins with a straight chain, beginning with propylene, are oligomerized in the presence of alkylaluminum compounds, while branched chain α-olefins are practically inert to these compounds. The one-step direct synthesis of alkylaluminum compounds involving branched chain α-olefins is technologically very simple, and side reactions which may occur during the synthesis of these compounds from aluminum, hydrogen and ethylene, are here absent. It was also noted that branched chain olefins such as isobutylene are not only inert to alkylaluminum compounds, but are also less readily hydrogenated than ethylene or propylene.

Two-stage direct synthesis of triethylaluminum

The formation of triethylaluminum from aluminum, hydrogen and ethylene comprises two stages: diethylaluminum hydride is formed first, and is then ethylated to triethylaluminum. Two moles of initial triethylaluminum will yield only one mole of this product.

Synthesis of diethylaluminum hydride

The study of the conditions of the reaction

$$Al + 2Al(C_2H_5)_3 + 1^1/_2H_2 \longrightarrow 3Al(C_2H_5)_2H$$

known as the hydrogenation step, was affected in two stages.

At first the most rational mode of preparation of active aluminum powder was selected /203 — 211/; subsequently, the realization of this process was studied under pressures of 20 — 50 atm, instead of pressures of the order of 100 — 150 atm, which had been formerly proposed.

It was found as a result of these studies that, out of the large number of compounds tested for their catalytic effect on the reaction, titanium hydride /212, 213/ proved to be one of the best. In the presence of this catalyst, the conversion of aluminum was 90% at 110°C and under 30 atm pressure. In the absence of a catalyst hydrogenation failed to take place under these conditions. Both the degree of conversion of the aluminum and the reaction rate increased when the concentration of titanium hydride was increased.

A study of the effect of the main technological parameters on the conversion of aluminum and the reaction rate in the presence of this catalyst showed that the interaction of aluminum, hydrogen and triethylaluminum is highly sensitive to the temperature. Below 70°C the reaction fails to take place. The synthesis takes place at a significant rate only at 80°C. At 115°C the degree of aluminum conversion is at its peak, and neither the yield nor the rate of the synthesis become any larger when the temperature is raised beyond this point. At 160°C and above the reaction no longer takes place, while above 170°C the initial triethylaluminum undergoes intensive decomposition. This decomposition is irreversible.

It was also shown that the main reaction was accompanied by the formation of ethylene, ethane, butylene, butane, and butylethylaluminum hydride. Below 120°C the amount of the side products does not exceed 0.2%. Between 120 and 140°C the yield of the side products becomes significant, while above 170°C it becomes by far the main reaction /209/.

The formation of olefins, lower alkanes and mixed organoaluminum compounds is the result of complex processes of hydrogenation, dissociation, chain extension, and substitution, which may be written as follows:

$$Al(C_2H_5)_3 + H_2 \longrightarrow HAl(C_2H_5)_2 + C_2H_6$$
$$Al(C_2H_5)_3 \longrightarrow HAl(C_2H_5)_2 + C_2H_4$$
$$Al(C_2H_5)_3 + C_2H_4 \longrightarrow Al(C_2H_5)_2(C_4H_9)$$
$$Al(C_2H_5)_2(C_4H_9) + C_2H_4 \longrightarrow Al(C_2H_5)_3 + C_4H_8$$
$$Al(C_2H_5)_2(C_4H_9) + H_2 \longrightarrow HAl(C_2H_5)_2 + C_4H_{10}$$

Ziegler /26/ and Zakharkin et al. /202/ also pointed out that such reactions may occur.

The study of the synthesis of diethylaluminum hydride showed that an increase in the pressure from 30 to 105 atm resulted in a doubling of the reaction rate (from 0.152 to 0.37 gram-mole of $HAl(C_2H_5)_2$ per gram-atom of Al per hour). It also resulted in an increased content of ethane in the gaseous reaction products (from 3 to 18.4 vol. %), which means that the rate of the reaction between triethylaluminum and hydrogen is pressure-dependent.

In the preparation of diethylaluminum hydride according to

$$Al + 2Al(C_2H_5)_3 + 1^1/_2H_2 \longrightarrow 3HAl(C_2H_5)_2$$

triethylaluminum acts as the source of organic free radicals during the formation of diethylaluminum hydride from aluminum and hydrogen. Accordingly, the rate of formation of the hydride will depend on the concentration of triethylaluminum in the reaction mixture. In accordance with the equation, the stoichiometric ratio between triethylaluminum and aluminum should not be less than two. However, since the side reactions

$$Al(C_2H_5)_3 + H_2 \longrightarrow Al(C_2H_5)_2H + C_2H_6$$
$$Al(C_2H_5)_3 \longrightarrow Al(C_2H_5)_2H + C_2H_4$$

also take place and increase the expenditure of the initial triethylaluminum, the molar ratio usually employed is three. That this procedure is correct has been confirmed by studies of the effect of the molar ratio on the conversion of the aluminum and on the rate of formation of the hydride.

Reaction of mixtures of diethylaluminum hydride and triethylaluminum with ethylene

The products of the reaction between hydrogen, aluminum and triethylaluminum are a complex mixture which consists of the associates $[Al(C_2H_5)_3]_2$, $[HAl(C_2H_5)_2]_3$, $HAl(C_2H_5)_2 \cdot Al(C_2H_5)_3$, etc. When this mixture is acted upon by ethylene, triethylaluminum is eventually formed:

$$[HAl(C_2H_5)_2]_3 \underset{}{\overset{C_2H_4}{\rightleftarrows}} \{[HAl(C_2H_5)_2]_2 \cdot Al(C_2H_5)_3\} \overset{C_2H_4}{\rightleftarrows}$$

$$\rightleftarrows [HAl(C_2H_5)_2 \cdot Al(C_2H_5)_3 + Al(C_2H_5)_3] \overset{C_2H_4}{\rightleftarrows} 1^1/_2[Al(C_2H_5)_3]_2$$

These reactions may also be accompanied by side processes, such as the formation of higher organoaluminum compounds:

$$Al(C_2H_5)_3 + (m+n+p)C_2H_4 \longrightarrow Al\begin{cases}(C_2H_4)_mC_2H_5\\(C_2H_4)_nC_2H_5\\(C_2H_4)_pC_2H_5\end{cases}$$

or by the formation of higher olefins

$$Al\begin{cases}(C_2H_4)_mC_2H_5\\(C_2H_4)_nC_2H_5\\(C_2H_4)_pC_2H_5\end{cases} + 3C_2H_4 \longrightarrow mCH_2{=}CH(C_2H_4)_{m-1}C_2H_5 +$$

$$+ nCH_2{=}CH(C_2H_4)_{n-1}C_2H_5 + pCH_2{=}CH(C_2H_4)_{p-1}C_2H_5 + Al(C_2H_5)_3$$

It has been experimentally shown /209/ that the course of the reaction between ethylene, diethylaluminum hydride and triethylaluminum is significantly affected by the temperature, the pressure of ethylene and the duration of the reaction.

The conversion of diethylaluminum hydride increases considerably with the temperature. When the reaction lasts for 30 minutes, the conversion of diethylaluminum hydride is 45% at 60°C, but is as much as 74% at 100°C. A complete conversion of diethylaluminum hydride under 10 atm pressure at 75°C was attained within $2^1/_2$ hours; at 100°C under 20 atm pressure it was attained after as little as 1.2 hours. It must be noted that when the temperature was raised to 115°C under 20 atm pressure, the conversion did not significantly increase, but the rate of absorption of the ethylene increased, which means that side reactions take place under these conditions.

An analysis of the composition of the reaction products confirmed that a side process does indeed take place. In fact, at 10 atm and 75°C diethylaluminum hydride is fully converted within $2^1/_2$ hours, mainly to triethylaluminum, up to 5% diethylbutylaluminum being formed as the side product. Relatively minor changes in the experimental conditions, such as increasing the temperature to 100°C or increasing the pressure to 20 atm, result in major changes in the composition of the reaction product: the yield and the content of triethylaluminum decrease, while those of diethylaluminum increase. At 20 atm, 115°C and reaction duration of 1.2 hours diethylaluminum becomes the main reaction product, and accounts for more than 50% of the total product. The rate of formation of triethylaluminum from diethylaluminum hydride present in concentrations of 30 to 50 mole % was very fast at first (up to 3.8 gram-moles $Al(C_2H_5)_3$ per gram-atom of $HAl(C_2H_5)_2$ per hour), but rapidly decreased to zero within 0.5 to 1.5 hours.

It is seen that the ethylation of mixtures of diethylaluminum hydride with triethylaluminum should not be carried out above 75°C and not above 20 atm pressure. When these parameters are increased, diethylaluminum is produced at a rapid rate.

One-step synthesis of triethylaluminum from hydrogen, aluminum and ethylene

The experimental conditions employed in the one-step synthesis of triethylaluminum favor both the main process and the side reactions (Table 2):

$$Al + 2Al(C_2H_5)_3 + 1^1/_2H_2 \xrightarrow[20-300\ atm]{100-140°\ C} 3Al(C_2H_5)_2H \xrightarrow[\sim 10\ atm,\ C_2H_4]{\sim 70\ °C} 3Al(C_2H_5)_3$$

TABLE 2. Main and side reactions which may take place during a one-step synthesis of triethylaluminum

Reaction	Conditions under which reaction is possible		References
	temperature, °C	pressure, atm	
Synthesis of diethylaluminum hydride $Al + 1^1/_2H_2 + 2Al(C_2H_5)_3 \longrightarrow 3HAl(C_2H_5)_2$	100 — 140	15 — 300	196, 201, 202
Synthesis of triethylaluminum $3HAl(C_2H_5)_2 + 3C_2H_4 \longrightarrow 3Al(C_2H_5)_3$	60 — 70	Up to 20	202
Synthesis of higher alkylaluminum compounds $(C_2H_5)_3Al + (m+n+p)C_2H_4 \longrightarrow Al\begin{cases}(C_2H_4)_mC_2H_5\\(C_2H_4)_nC_2H_5\\(C_2H_4)_pC_2H_5\end{cases}$	90 — 120	Not less than 80	26
Formation of higher olefins $Al\begin{cases}(C_2H_4)_mC_2H_5\\(C_2H_4)_nC_2H_5\\(C_2H_4)_pC_2H_5\end{cases} + 3C_2H_4 \longrightarrow CH_2{=}CH(C_2H_4)_{m-1}C_2H_5 + CH_2{=}CH(C_2H_4)_{n-1}C_2H_5 + CH_2{=}CH(C_2H_4)_{p-1}C_2H_5 + Al(C_2H_5)_3$	1 — 100	1 — 100	26
Hydrogenation of triethylaluminum $Al(C_2H_5)_3 + H_2 \longrightarrow HAl(C_2H_5)_3 + C_2H_6$	140 — 160	Elevated	26
Radical exchange $AlR_3 + AlR'_3 + AlR''_3 \longrightarrow 3AlRR'R''$	20		26
Hydrogenation of ethylene $C_2H_4 + H_2 \longrightarrow C_2H_6$	85 — 135	Up to 100	200

It was accordingly very difficult to find experimental conditions under which the reaction between hydrogen, aluminum and ethylene would stop at the triethylaluminum stage, without formation of higher alkylaluminum compounds or some other products (Table 2).

Repeated attempts finally resulted in a product which contained practically only triethylaluminum /214/. The synthesis could be realized under a low partial pressure of ethylene and an ethylene:hydrogen ratio lower than the stoichiometric. The overall pressure under which the process is realized is 50 atm.

It was shown that if the initial ethylene:hydrogen ratio is 2:1 and the reaction temperature is 100 — 135°C, the reaction products are higher alkylaluminum compounds, the formation of which consumes between 3 and 49% of aluminum and between 23.6 and 49.1% of triethylaluminum. It is only at 135°C that no conversion of triethylaluminum into higher alkylaluminum compounds was observed. It was also noted that the temperature of the synthesis determines the composition of the reaction products. Thus, for instance, in a synthesis conducted at 100°C triethylaluminum, tributylaluminum and trihexylaluminum are all formed; at 120°C higher alkylaluminum compounds up to tridecylaluminum are produced; if the reaction is conducted at 150°C, higher olefins are obtained. A study of the effect of the molar ratio $Al:Al(C_2H_5)_3$ showed that the amount of triethylaluminum increases only if the value of this ratio is 1: 0.71. Under any other conditions the newly formed triethylaluminum is converted to higher alkylaluminum compounds.

A study of the effect of the molar ratio between ethylene and hydrogen showed that, when this ratio is decreased from 2:1 to 1:1, at 135°C and under 50 atm pressure, the conversion of aluminum increases from 49 to 67%, and the rate of formation of triethylaluminum increases from 0.00215 to 0.0515 gram-mole $Al(C_2H_5)_3$ per gram-atom of aluminum per hour. The content of triethylaluminum in the reaction products increases from 61.2 to 89.0 mole %. If the $C_2H_4:H_2$ ratio is further decreased to 0.5:1, the increase in the triethylaluminum content is insignificant (to 90.7%), but the aluminum conversion decreases to 44%. Accordingly, the optimum value of the ethylene:hydrogen ratio during the one-step synthesis is 1:1 rather than 0.5:1.

Synthesis of diisobutylaluminum hydride and of triisobutylaluminum

The synthesis of triisobutylaluminum can be effected both in one and two steps, and at a faster rate than the synthesis of triethylaluminum. Isobutyl derivatives of aluminum are prepared under 50 — 60 atm pressure and at elevated temperatures:

$$Al + 1^1/_2H_2 + 2Al(\text{iso-}C_4H_9)_3 \longrightarrow 3Al(\text{iso-}C_4H_9)_2H$$
$$3Al(\text{iso-}C_4H_9)_2H + 3\ \text{iso-}C_4H_8 \longrightarrow 3Al(\text{iso-}C_4H_9)_3$$
$$\overline{Al + 1^1/_2H_2 + 3\ \text{iso-}C_4H_8 \longrightarrow Al(\text{iso-}C_4H_9)_3}$$

Studies of one-step synthesis of triisobutylaluminum showed /215/ that the effect of pressure on the course of the one-step synthesis of triisobutylaluminum is considerable only up to a pressure of 60 atm, the conversion of aluminum rapidly increasing from 12.3% when the synthesis is carried out under 30 atm pressure to 71% when it is carried out under 60 atm pressure. Further increase in pressure does not materially affect the degree of conversion of aluminum.

A high conversion of aluminum (up to 73.2%) and a reaction rate of 0.12 g-mole $Al(iso\text{-}C_4H_9)_3$ per g-atom Al per hour is attained at 150°C. A decrease in the temperature results in a decreased degree of conversion of aluminum. Above 150°C the reaction rate decreases owing to the increase in the rate of decomposition of triisobutylaluminum. Varying the initial ratio $Al(iso\text{-}C_4H_9)_3$: Al between 0.17 and 2.79 did not significantly alter the results of the synthesis. Both the reaction rate and the conversion of aluminum remained constant when the value of this ratio was varied.

It is assumed that in the immediate vicinity of the aluminum surface triisobutylaluminum acts as carrier of alkyl radicals to the aluminum and its diffusion toward the reaction surface in the course of the synthesis is not the rate-determining factor. The initial ratio of isobutylene to aluminum is on the other hand a major factor in the conversion of aluminum and in the rate of synthesis of triisobutylaluminum. In all cases (trials under 30 and 60 atm pressure), a decrease in this ratio resulted in a decreased reaction rate, and also in a decreased conversion of aluminum. It was found /215/ that the optimum isobutylene : aluminum ratio to be employed in the synthesis is 6.0.

The synthesis of diisobutylaluminum hydride is usually effected by the reaction:

$$Al + 2Al(iso\text{-}C_4H_9)_3 + 1^1/_2H_2 \longrightarrow 3Al(iso\text{-}C_4H_9)_2H$$

Similarly to the one-step synthesis of triisobutylaluminum, this reaction takes place at 40 — 150°C and under 50 — 60 atm pressure /210/. Raising the pressure to 100 atm does not materially affect the parameters of the synthesis. If the initial ratio $Al(iso\text{-}C_4H_9)_3$: Al is close to two, the reaction rate is 0.107 gram-mole $HAl(iso\text{-}C_4H_9)_2$ per g-atom Al per hour, and the conversion of aluminum is 64.5%. If the $Al(C_4H_9)_3$: Al ratio is increased to 3 : 1, the rate of the synthesis rapidly increases, while the conversion of the aluminum attains 98%. If the ratio is reduced to 1.76, the reaction rate and the conversion of aluminum both decrease considerably, which means that the synthesis of isobutylaluminum dihydride

$$2Al + Al(iso\text{-}C_4H_9)_3 + 3H_2 \longrightarrow 3H_2Al(iso\text{-}C_4H_9)$$

cannot proceed under these conditions.

The optimum value of the initial ratio $Al(iso\text{-}C_4H_9)_3$: Al is assumed to be three, when a satisfactory reaction rate and degree of conversion of the aluminum can be attained.

Synthesis of higher alkylaluminum compounds

One-step syntheses of triethylaluminum and triisobutylaluminum may also be used to prepare other organoaluminum compounds. However, the optimum conditions for such reactions have not yet been determined with certainty; this applies in particular to the preparation of alkylaluminum compounds with straight chain hydrocarbon radicals. It is preferable to prepare higher alkylaluminum compounds in two steps as described above. The first step should be realized at 110 — 140°C and under 50 — 200 atm pressure:

$$Al + 2Al(C_8H_{17})_3 + 1^1/_2H_2 \longrightarrow 3Al(C_8H_{17})_2H$$

while the second takes place at 70 — 90°C in the absence of hydrogen:

$$3Al(C_8H_{17})_2H + 3C_8H_{16} \longrightarrow 3Al(C_8H_{17})_3$$

Clearly, either trialkylaluminum compounds or dialkylaluminum hydrides may be prepared in this way, by arresting the reaction at the appropriate stage. In the latter case, some impurity trialkylaluminum remains behind in the reaction mixture. This product can be converted to dialkylaluminum hydride by heating the reaction mixture to 140 — 160°C in the presence of hydrogen.

Another method for the preparation of alkylaluminum compounds, which is particularly convenient in small-scale work, is the direct interaction of the olefin with triisobutylaluminum:

$$Al(\text{iso-}C_4H_9)_3 + 3C_nH_{2n} \longrightarrow Al(C_nH_{2n+1})_3 + 3\ \text{iso-}C_4H_8$$

In performing this reaction, the following factors must be taken into account /26/: 1) if the boiling point of the olefin component is higher than that of isobutylene, the latter can be removed by evaporation; 2) if the olefin boils at the boiling point of isobutylene ($CH_3CH_2CH = CH_2$) or below it, isobutylene can still be removed, since α-olefins have a stronger affinity to compounds with a $>$Al — H bond than has isobutylene; 3) the reaction must be initiated above 100°C; on the other hand, when working with α-olefins of the type $RCH = CH_2$, the reaction temperature should be slightly above 110°C, to prevent catalytic oligomerization of the olefin.

The most convenient variety of synthesis of alkylaluminum compounds from triisobutylaluminum and olefins is the method of Larbig /26/, in which isobutylene or triisobutylaluminum is first split off at about 100°C; this is followed by the addition of an equivalent amount of olefin. This process can be conducted continuously at a rate which ensures the presence of free alkylaluminum hydride in the reaction mixture, by varying the amount of the cleaved isobutylene. As soon as all isobutylene has distilled off, the remainder of the olefin is introduced, the reaction mixture is cooled to 60°C and left to stand for several hours. In this way practically pure trialkylaluminum compounds, free from dimerization products, are obtained.

Alkylaluminum compounds can be prepared in this way not only from olefins of the types $RCH = CH_2$ and $RR'C = CH_2$, but also from vinylcyclohexene, limonene, β-pinene, camphene, etc. In individual cases (e.g., camphene) dialkylaluminum hydride is the main product. If hydrocarbons with a double bond in the middle of the chain are employed, the reaction will proceed readily only in the presence of catalysts — salts of transition metals /216/. Information is also available on the reaction between triethylaluminum and various olefins /102/.

Higher alkylaluminum compounds can also be prepared by chain extension reaction resulting from the interaction of an alkylaluminum compound with ethylene /26, 97/:

$$AlC_2H_5 + nC_2H_4 \longrightarrow Al(C_2H_4)_nC_2H_5$$

The scope of the reaction is limited by the fact that when α-branched alkylaluminum compounds are employed, displacement of the olefin is the preferred reaction.

The reaction of chain extension can only be successfully performed if a number of its characteristic features are allowed for. In the first place, it only takes place if all three valencies of the aluminum atom are bound to alkyl groups. Trimethylaluminum, which does not react with ethylene, is the only exception to this rule. It was shown by Ziegler et al. /26/ that the reaction must be performed at 90 — 120°C and that the pressure of ethylene must be at least 60 atm. The rate of addition of ethylene at 95 — 105°C and under 80 — 90 atm pressure is on the average 1 mole per 1 mole of alkylaluminum compound per hour.* If the temperature is raised, the reaction rate increases, and the entire course of the reaction is changed. The heat of reaction, which is practically equal to the heat of polymerization of ethylene (about 22 kcal/mole) is evolved all at once, with the result that the rise in the temperature is irregular, and the reaction often terminates in a flash, which is accompanied by a complete decomposition of the ethylene into methane, hydrogen and carbon. According to Ziegler et al., explosions in small-sized laboratory autoclaves will take place above 125°C and 125 atm pressure. This danger is not so great if the starting material consists of higher alkylaluminum compounds or if the triethylaluminum is diluted with saturated hydrocarbons.

An increase in the pressure of ethylene and in the temperature also increases the rate of the displacement reaction, resulting in the elimination of an olefin:

$$R(CH_2-CH_2)_n\,al + C_2H_4 \rightleftarrows R(CH_2-CH_2)_{n-1}-CH=CH_2 + C_2H_5\,al$$

This reaction also takes place in the presence of traces of metals such as nickel, especially in the colloidal form. The olefins evolved dimerize at the same time.

All Al — C bonds present in the reaction mixture at any given moment of time have the same probability of adding on to the new ethylene molecule. The distribution of the reaction products may be approximately represented by the expression

$$x_{(p)} = \frac{n^p e^{-n}}{p!}$$

where $x_{(p)}$ is the molar ratio of chains with p added ethylene molecules, n is the number of moles of ethylene consumed in accordance with the equation

$$R-Al + nC_2H_4 \longrightarrow R(C_2H_4)_n\,Al$$

* Recently /26 a, 26 b/ this reaction was successfully performed on dialkylaluminum chloride (in lieu of trialkylaluminum) at a lower pressure.

per equivalent R — Al; p is the number of ethylene molecules in individual hydrocarbon chains.

It has been noted that the rate of absorption of ethylene by triethylaluminum per mole of the alkylaluminum compound strongly increases with dilution.

In the addition of alkylaluminum compounds with a straight hydrocarbon chain to ethylene it is immaterial whether the chain contains an odd or an even number of carbon atoms. Reactions between ethylene and alkylaluminum compounds with an odd number of carbon atoms in the hydrocarbon group are distinguished by two typical features. Firstly, there is no significant reaction between trimethylaluminum and ethylene under conditions considered as the optimum conditions for the reactions between alkylaluminum compounds and ethylene. Such reaction requires more drastic conditions. Secondly, higher alkylaluminum compounds with an odd number of carbon atoms cannot be prepared by this method. This is because the side reaction of the displacement of olefin with an odd number of carbon atoms by ethylene (which invariably takes place) results in the formation of triethylaluminum and subsequently in the formation of higher alkylaluminum compounds with an even number of carbon atoms.

5. PREPARATION OF ALKYLALUMINUM HALIDES

Synthesis of alkylaluminum sesquihalides

We have already said that one convenient method of preparing alkylaluminum halides is by the reaction between aluminum and alkyl halides:

$$2Al + 3RX \longrightarrow AlR_2X + AlRX_2$$

Aluminum reacts with alkyl chlorides, alkyl bromides and alkyl iodides in different manners /86/. Thus, reactions with alkyl halides RX where X is Br or I and R is methyl, ethyl or propyl proceed under mild conditions in high yields and may be applied in the laboratory. However, even butyl bromide reacts very slowly and the yield of the final product is small. Attempts to prepare alkylaluminum sesquihalides by reaction with pentyl bromide were unsuccessful; isopentyl and octylaluminum sesquibromides have been obtained only recently /1/. Aluminum reacts with diarylhalides under mild conditions; this particularly applies to iodides /86/.

Reactions between aluminum and alkyl chlorides proved to be the most difficult to realize. These syntheses are usually performed in metal vessels designed to withstand high pressures. The reaction between aluminum and ethyl chloride, without special activation, at first proceeds sluggishly. According to Kryukov et al. /217/, the induction period may last for several days. Once the reaction has started, it proceeds very rapidly and is difficult to regulate. The use of primers (iodine, bromine, aluminum chloride, etc.) makes it possible to reduce the induction period of the reaction, but this effect is insufficient for practical purposes. It was shown by Zhigach et al. /218/ that if aluminum is previously activated with ethyl bromide, and the reaction is conducted in the presence of hydrocarbons, the formation of ethylaluminum sesquichloride is not preceded by an induction period; however, tars were formed in the reactor in a number of cases. Better results were obtained by fully immersing the aluminum in the ethylaluminum sesquichloride medium and maintaining a continuous supply of ethyl chloride to the reaction mixture. This process also proved unsuitable for industrial application, due to certain difficulties involved in the introduction of aluminum powder into the reaction mixture.

At a later date Korneev, Popov et al. /219/ established the optimum conditions for the reaction between aluminum and ethyl chloride in a solution of ethylaluminum sesquichloride in petroleum ether. It was found that the process could be realized with various kinds of aluminum, but that it was best to use aluminum powder manufactured in the Soviet Union, of 10 — 50 μ particle size. This aluminum can be activated by keeping for two hours at 50 — 60°C in ethylaluminum sesquihalide medium. The activation temperature is maintained at the desired level by a gradual addition of ethyl bromide (0.25 kg C_2H_5Br is taken for each 0.61 kg aluminum). After such an activation aluminum readily reacts with ethyl chloride, without an induction period. The introduction of iodine into the reaction mixture does not materially affect the results of the synthesis. The reaction temperature has a major effect on the conversion of aluminum and the reaction rate. The best results were obtained at 135°C; at higher temperatures butane is formed in a side reaction:

$$Al(C_2H_5)_2Cl + C_2H_5Cl \longrightarrow Al(C_2H_5)Cl_2 + C_4H_{10}$$

At the same time the exothermal effect rapidly increases, and the reaction mixture contains a large amount of tars. According to the authors it is necessary to employ high-speed paddle stirrers in the process.

In another publication /220/ it is recommended to conduct the reaction in a continuous process reactor with an external cooler to carry away the heat of the reaction.

Synthesis of dialkylaluminum halides

Several methods may be employed to prepare dialkylaluminum halides from alkylaluminum sesquihalides. The simplest method is to use the reaction between alkylaluminum sesquihalide and trialkylaluminum /38, 209/:

$$Al_2R_3X_3 + AlR_3 \longrightarrow 3AlR_2X$$

This reaction begins to take place at room temperature and readily proceeds with an evolution of heat, which is very marked if ethyl or methyl derivatives of aluminum are employed.

It is also possible to prepare dialkylaluminum halides, in particular diethylaluminum chloride, by the reaction /221/:

$$Al_2(C_2H_5)_3Cl_3 + NaCl \longrightarrow Al(C_2H_5)_2Cl + NaAl(C_2H_5)Cl_3$$

The reaction takes place at a high temperature and in the presence of a large excess of sodium chloride. The sodium chloride should be carefully dried prior to use. In this way the diethylaluminum chloride product is practically free from ethylaluminum dichloride.

Another simple synthesis of dialkylaluminum halides is the reaction between the appropriate trialkylaluminum compound and aluminum halide /38, 209/:

$$2AlR_3 + AlX_3 \longrightarrow 3AlR_2X$$

Dialkylaluminum bromides and iodides can readily be obtained in this way. The preparation of dialkylaluminum chloride proceeds in a somewhat different manner, owing to the poor solubility of aluminum trichloride in hydrocarbons. The reaction between trialkylaluminum and aluminum trichloride is a case of conversion of more alkylated to less alkylated organoaluminum compounds by reacting the former with aluminum alkylates or aluminum halides. Despite the apparent simplicity of the reaction, the reaction of trialkylaluminum compounds with $AlCl_3$ or even with alkylaluminum sesquichloride is accompanied by side reactions, and the final products are not pure. The reaction is accompanied by the evolution of gases and precipitation of metallic aluminum /222/. The side reactions were due to the thermal decomposition of the reaction products, since the alkylaluminum compounds were prepared at 140 — 190°C. The study of this reaction, carried out by Korneev et al. /223/, showed that dialkylaluminum chloride free from side products is obtained at a temperature not above 70°C. Data on the pyrolysis of alkylaluminum compounds indicate that the formation of such side products may be represented as follows:

$$>AlC_2H_5 \longrightarrow >AlH + CH_2{=}CH_2$$

$$>AlC_2H_5 + CH_2{=}CH_2 \longrightarrow >AlC_4H_9$$

$$>AlC_4H_9 + CH_2{=}CH_2 \longrightarrow >AlC_2H_5 + C_4H_8$$

$$3(C_2H_5)_2AlH + AlCl_3 \longrightarrow 3(C_2H_5)_2AlCl + AlH_3$$

$$AlH_3 \longrightarrow Al + 1^1/_2H_2$$

$$H_2 + C_4H_8 \longrightarrow C_4H_{10}$$

$$H_2 + C_2H_4 \longrightarrow C_2H_6$$

The determination of the rates of formation of diethyl- and diisobutyl-aluminum chloride as a function of temperature showed that these processes take place at measurable rates at low temperatures (−20 to +20°C) and become greatly accelerated when the temperature is raised to 50 — 80°C. When preparing dialkylaluminum chlorides, the reaction mixture must be thoroughly stirred and the solution of trialkylaluminum should be introduced into a 20 — 30% suspension of aluminum chloride. It was shown by Zhigach et al. /224/ that diethylaluminum chloride can also be prepared by reacting aluminum and magnesium powders with ethyl chloride. The reaction proceeds readily in hydrocarbon medium with up to 75% yield at 70 — 100°C.

The preparation of diethylaluminum chloride by dehalogenation of ethyl-aluminum sesquichloride by metallic sodium is the only reaction employed in industry /219, 225/. The reaction may be represented as follows /38/:

$$2Al_2(C_2H_5)_3X_3 + 3Na \longrightarrow 3Al(C_2H_5)_2X + 3NaX + Al$$
(where X = Cl, Br, I)

The study of this reaction, which was performed in order to establish the optimum parameters of the synthesis and to obtain purer polymerization products, revealed that dehalogenation involves a series of chemical reactions. It was noted that the reaction can be realized in two ways. In the first variant, molten metallic sodium is fed to a hydrocarbon solution of ethylaluminum sesquichloride ("sodium — ethylaluminum sesquichloride method"), while in the second a hydrocarbon solution of $Al_2(C_2H_5)_3Cl_3$ is fed into an emulsion of sodium in the hydrocarbon ("ethylaluminum sesquichloride — sodium method"). In the former case the reaction takes place in excess $Al_2(C_2H_5)_3Cl_3$, while in the latter it takes place in excess sodium. Thus, depending on the variant chosen, the dehalogenation reaction involves the formation of the complexes $NaAl(C_2H_5)Cl_3$ or $NaAl(C_2H_5)_4$.

It has been found that when the reaction is realized by the "sodium — ethylaluminum sesquichloride" method, the reaction initially proceeds as follows:

$$2Al_2(C_2H_5)_3Cl_3 + 3Na \longrightarrow 3Al(C_2H_5)_2Cl + 3NaCl + Al$$

The ethylaluminum chloride present in the reaction mixture reacts with the NaCl formed:

$$NaCl + Al(C_2H_5)Cl_2 \longrightarrow NaAl(C_2H_5)Cl_3$$

Thus, ethylaluminum sesquichloride is fully expended even when the amount of the sodium which is fed in is much smaller. The overall reaction may be written as follows:

$$3Na + 5Al_2(C_2H_5)_3Cl_3 \longrightarrow 6Al(C_2H_5)_2Cl + 3NaAl(C_2H_5)Cl_3 + Al$$

On the addition of sodium the reaction proceeds in two ways /38/:

$$3Na + 2NaAl(C_2H_5)Cl_3 \longrightarrow Al(C_2H_5)_2Cl + 5NaCl + Al$$

and

$$3Na + 3Al(C_2H_5)_2Cl \longrightarrow 2Al(C_2H_5)_3 + 3NaCl + Al$$

The triethylaluminum formed reacts with sodium aluminum ethyl trichloride:

$$NaAl(C_2H_5)Cl_3 + Al(C_2H_5)_3 \longrightarrow 2Al(C_2H_5)_2Cl + NaCl$$

and also with sodium:

$$4Al(C_2H_5)_3 + 3Na \longrightarrow 3NaAl(C_2H_5)_4 + Al$$

The dehalogenation process according to the "ethylaluminum sesquichloride — sodium method" proceeds in a different manner, viz.:

$$4Al_2(C_2H_5)_3Cl_3 + 15Na \longrightarrow 3NaAl(C_2H_5)_4 + 12NaCl + 5Al$$

$$2Al_2(C_2H_5)_3Cl_3 + NaAl(C_2H_5)_4 \longrightarrow 5Al(C_2H_5)_2Cl + NaCl$$

$$Al_2(C_2H_5)_3Cl_3 + NaCl \longrightarrow Al(C_2H_5)_2Cl + NaAl(C_2H_5)Cl_3$$

$$Al(C_2H_5)_2Cl + NaAl(C_2H_5)_4 \longrightarrow 2Al(C_2H_5)_3 + NaCl$$

$$NaAl(C_2H_5)Cl_3 + Al(C_2H_5)_3 \longrightarrow 2Al(C_2H_5)_2Cl + NaCl$$

i.e., $NaAl(C_2H_5)_4$ is formed first and subsequently reacts with $Al_2(C_2H_5)_3Cl_3$ up to the formation of diethylaluminum chloride.

It is interesting to note that when this variant is employed, the liquid reaction products will contain diethylaluminum chloride almost exclusively, even if ethylaluminum sesquichloride is introduced in excess. This is because the excess $Al_2(C_2H_5)_3Cl_3$ reacts with sodium chloride with the formation of $NaAl(C_2H_5)Cl_3$, and only after all the NaCl in the solid reaction products has been consumed, can ethylaluminum dichloride appear in the liquid products along with diethylaluminum chloride. This assumption has been experimentally confirmed.

It was found as a result of a study of the temperature dependence of the yield of diethylaluminum chloride, of the reaction rate and of the composition of the reaction products, that the synthesis is best performed at 105 — 135°C. Above 160°C the yield and rate of formation of $Al(C_2H_5)_2Cl$ decrease, while the content of triethylaluminum in the reaction products increases.

Very probably, reactions between other alkylaluminum sesquihalides and metallic sodium involve the same stages, except for the compounds which do not form complexes with sodium halide.

Bibliography

1. Zhigach, A.F. and D.S. Stasinevich. Reaktsii i metody issledovaniya organicheskikh soedinenii (Reactions and Methods of Analysis of Organic Compounds). — Sb. 10:209. 1961.
2. Strohmeier, W. and K. Humpfner. — Chem. Ber., **90**:2339. 1957.
3. Pozamantir, A.G. and M.L. Genusov. — ZhOKh, **32**:1175. 1962.
4. Milovskaya, E.B., B.A. Dolgoplosk, and P.I. Dolgopol'skaya. — VMS, **10**:1503. 1962.
5. Eden, C. and H. Feilchenfeld. — J. Phys. Chem., **7**:1354. 1962.
6. Grunewald, G. — Ber., **21**:881. 1881.
7. Lonise, E. and L. Roux. — Bull. Soc. chim. France, **50**:497. 1888.
8. Laubengauer, A. and G. Gillam. — J. Am. Chem. Soc., **63**:477. 1941.
9. Pitzer, K. and H. Gutowsky. — J. Am. Chem. Soc., **68**:2204. 1946.
10. Nesmeyanov, A.N., O.V. Nogina, and V.A. Dubovitskii. — Doklady AN SSSR, **134**(6):1363. 1960.
11. Hoffman, E. — Lieb. Ann. Chem., **629**:104. 1960.
12. Patat, F. and H. Sinn. — Angew. Chem., **70**:496. 1958.
13. Coates, G. Organometallic Compounds, 2nd ed. — New York, Barnes and Noble. 1960.
14. Rundle, R. — J. Am. Chem. Soc., **69**:1327. 1947.
15. Pitzer, K. and R. Sheline. — J. Chem. Phys., **16**:552. 1948.
16. Hoffman, E. — Z. anal. Chem., **170**:177. 1959.
17. Davidson, N. and H. Brown. — J. Am. Chem. Soc., **64**:316. 1942.

17a. Ryokichi, T. — Bull. Chem. Soc. Japan, **39**(4):725. 1966.
18. Ziegler, K. and R. Köster. — Ann., **608**:1. 1957.
18a. Nesmeyanov, A. N. and R. A. Sokolik. Metody elementorganicheskoi khimii, B, Al, Ga, In, Tl, (Methods of Organometallic Chemistry, B, Al, Ga, In, Tl), p. 288. — Izd. "Nauka." 1964.
18b. Takeda, S. and R. Tarao. — Bull. Chem. Soc. Japan, **38**:1567. 1965.
19. Hoffman, E. — Z. Elektrochem., **61**:1014. 1957.
20. Strohmeier, W. and K. Hümpfner. — Z. Electrochem., **61**:1010. 1957.
21. Bonitz, E. — Chem. Ber., **88**:742. 1955.
22. Zeiss, H. H. (editor). Organometallic Chemistry. — New York. Reinhold, 1960.
23. Graevskii, A. I. et al. — Doklady AN SSSR, **119**:101. 1958.
24. Knap, E. et al. — Ind. Eng. Chem., **5**:874. 1957.
25. Jeddanapalli, L. and C. Schubert. — J. Chem. Phys., **14**:1. 1946.
26. Zhigach, A. F. (editor). Alyuminiiorganicheskie soedineniya (Organoaluminum Compounds). — IL. 1962. [Collection of translated articles].
27. Larikov, E. I. et al. — Khimicheskaya Promyshlennost', **3**:171. 1964.
28. Zakharkin, L. I. and O. Yu. Okhlobystin. — Izvestiya AN SSSR, OKhN, p. 1279. 1958.
29. Natta, G. et al. — J. Am. Chem. Soc., **81**:2561. 1959.
30. Zakharkin, L. I. and L. A. Savina. — Izvestiya AN SSSR, OKhN, p. 444. 1959.
31. Rochow, E., D. Hurd, and R. Lewis. The Chemistry of Organometallic Compounds. — New York, Wiley. 1957.
32. Shaulov, Yu. Kh. et al. — ZhFKh, **38**(7):1779. 1964.
33. Baker, E. and H. Sisler. — J. Am. Chem. Soc., **75**:4828. 1953.
34. Bamford, C. et al. — J. Chem. Soc., p. 408. 1946.
35. Bleekrode, L. — Rec. trav. chim., **4**:80. 1885.
36. Ziegler, K. et al. — Lieb. Ann. Chem., **589**:119. 1954.
37. Hein, G. and H. Pauling. — Z. phys. Chem., **165A**:338. 1933.
38. Grosse, A. and J. Mavity. — J. Org. Chem., **5**:106. 1940.
39. Nobis, J. — Ind. Eng. Chem., **49**(12):44A. 1957.
40. Shaulov, Yu. Kh., G. O. Shmyreva, and V. S. Tubyanskaya. — ZhFKh, **39**:1. 1965.
41. Studiengesellschaft Kohle, Dutch Patent 976671. 1961; Austrian Patent 202566. 1959.
42. Ziegler, K. — French Patent 1291354. 1962; Austrian Patent 228229. 1963.
43. Ethyl Corporation, British Patent 897334. 1962; GFR Patent 1120448. 1962.
44. Ziegler, K. and H. G. Gellert. — GFR Patent 918928. 1955.
45. Robinson, R. — US Patent 2915542. 1959.
46. Hamprecht, G. and M. Schwarzmann. — British Patent 869179. 1961; GFR Patent 1122952. 1958.
47. Ziegler, K. — British Patent 923652. 1963; French Patent 1261147. 1961.
48. Ziegler, K. — Austrian Patent 224127. 1962; French Patent 1287337. 1962.
49. Hurd, D. — J. Org. Chem., **13**:771. 1948.
50. Baker, E. and H. Sisler. — J. Am. Chem. Soc., **75**:5193. 1953.
51. Podall, H. — Ethyl Corp., US Patent 3046290. 1962.

52. Ziegler, K. and K. Zosel. — US Patent 2691668. 1954.
53. Ziegler, K. and H. Martin. — Makromol. Chem., **18**:186. 1956.
54. Ziegler, K. and K. Zosel. — GFR Patent 916167. 1954.
55. Jenkner, H. — GFR Patent 1048581. 1959.
56. Ziegler, K. — British Patent 836734. 1960; Italian Patent 567577. 1957.
57. Continental Oil Co., British Patent 867986. 1961; French Patent 1246540. 1960.
58. Montecatini Società Generale per L'Industria Mineraria Chimica, British Patent 928716. 1963; Austrian Patent 228231, 228232. 1963.
59. Kali-Chemie, A.G., — French Patent 118181323. 1960.
60. Zakharkin, L.I. and O.Yu. Okhlobystin. — Soviet Patent 110977. 1957.
61. Zakharkin, L.I. and O.Yu. Okhlobystin. — Izvestiya AN SSSR, OKhN, **11**:1942. 1959; Soviet Patent 110674. 1957.
62. Jenkner, H. — Z. Naturforsch., **12**:809. 1957.
63. Barbaras, G. et al. — J. Am. Chem. Soc., **73**:4585. 1951.
64. Zakharkin, L.I. and I.M. Khorlina. — ZhOKh, **32**(9):2783. 1962.
65. Ziegler, K. and E. Holzkamp. — Lieb. Ann., **605**:93. 1957.
66. Bähr, G. and G. Müller. — Chem. Ber., **88**:251. 1951.
67. Ziegler, K. — French Patent 1321086. 1963.
68. Scherer, O. and G. Schalffer. — GFR Patent 1113218. 1963.
69. Nobis, J. — US Patent 3006942. 1961.
70. Jenkner, H. — GFR Patent 1009630. 1957.
71. Hamprecht, G. and M. Tittel. — GFR Patent 1116660. 1962.
72. Hamprecht, G. — GFR Patent 1102151. 1961.
73. Vranka, R. et al. — J. Am. Chem. Soc., **89**(13):3121. 1967.
74. Köster, R. and G. Benedikt. — Angew. Chem., **74**(15):589. 1962.
75. Koster, R. — GFR Patent 1057600. 1959; Köster, R. — Lieb. Ann., **618**:31. 1958.
76. Jenkner, H. — GFR Patent 1028576. 1958.
77. Köster, R. — Angew. Chem., **70**:371. 1958; Aschby, E. — J. Am. Chem. Soc., **81**:4791. 1959; Zakharkin, L.I. and O.Yu. Okhlobystin. — Izvestiya AN SSSR, OKhN, **6**:1135. 1959.
78. Jenkner, H. — GFR Patent 1028120. 1958; Ziegler, K. — Austrian Patent 228228. 1958; Hunt, M. — French Patent 1270629. 1961.
79. Jenkner, H. — GFR Patent 1121048. 1962; French Patent 1268177. 1961; Imp. Chem. Ind., Dutch Patent 101365. 1962; Austrian Patent 210436. 1960.
80. Siemens-Schuckertwerke, A.G. — French Patent 1294974. 1963.
81. Zakharkin, L.I. and V.V. Gavrilenko. — Izvestiya AN SSSR, OKhN, **1**:173. 1962.
82. Ziegler, K., R. Köster, and W. Kroll. — GFR Patent 1055534. 1959.
83. Organometalliques problèmes de structure et reactions nouvelles; Colloque international de Chimie organique, Paris 24 — 28 Septembre. 1962.
84. Nicolescu, I. et al. — Izvestiya AN SSSR, OKhN, **1**:94. 1960.
85. Groizeleau, L. — Compt. rend., **244**:1223. 1957.
86. Nesmeyanov, A.N. and R.A. Sokolik. Metody elementorganicheskoi khimii, B, Al, Ga, In, Tl (Methods of Organometallic Chemistry of B, Al, Ga, In, Tl). — Izd. "Nauka." 1964.

87. Golovanenko, B.I. and A.T.Menyailo. — Soviet Patent 132629. 1959.
88. Meiners, A. and F.Morris. — US Patent 2962513. 1960.
89. Labo, P. — US Patent 2971969. 1961; Ziegler, K. and P.Labo. — GFR Patent 1130808. 1962; French Patent 1251038. 1960.
90. Robinson, S. and E.Sidebottom. — French Patent 1324708. 1963.
91. Continental Oil Co., British Patent 883041. 1961.
92. Zosel, K. — GFR Patent 1104509. 1961.
93. Johnson, J. — US Patent 2863896. 1958; British Patent 837119. 1960; French Patent 1246124. 1960; Italian Patent 574936. 1958.
94. Ziegler, K. — US Patent 3013043. 1961; Ziegler, K. — French Patent 1226578. 1960.
95. Continental Oil Co., French Patent 1243143. 1960; British Patent 885612. 1961.
96. Gaylord, N. and H.Mark. Linear and Stereoregular Addition Polymers. — New York, Interscience. 1959.
97. Menyailo, A.T. et al. — Khimicheskaya Promyshlennost', **4**:11. 1965.
98. Ziegler, K. and H.Gellert. — Lieb.Ann.Chem., **589**:91. 1954.
99. Ziegler, K. — Angew.Chem., **64**:323. 1958.
100. Natta, G. et al. — J.Am.Chem.Soc., **81**(10):2561. 1959.
101. Pasynkiewiez, S. et al. — J.Organomet.Chem., **8**(2):233. 1967.
102. Allen, P. et al. — Trans.Faraday Soc., **63**(7):1636. 1967.
103. Henkel, H. and G.Gie. — British Patent 914053. 1962.
104. Asinger, F. et al. — Chem.Ber., **97**(9):2515. 1965.
105. Ashby, E. et al. — US Patent 3020298. 1962.
106. Petrov, A.A., V.S.Zavgorodnii, and V.A.Kormer. — ZhOKh, **32**(4):1349. 1962.
107. Ziegler, K. and H.Lehmkuhl. — Chem.Age, **81**:1047. 1959; Ziegler, K. — Soviet Patent 132136. 1959; French Patent 1300377. 1962.
108. Ethyl Corp., French Patent 1298231. 1962.
109. Nall, W. and H.Jones. — GFR Patent 888852. 1957; Farbwerke Hoechst, A.G. — British Patent 839370. 1960; Jenkner, H. — GFR Patent 1117577. 1962.
110. Ziegler, K. — GFR Patent 942026. 1957; British Patent 778098. 1957.
111. Ziegler, K. et al. — US Patent 3015669. 1962.
112. Ziegler, K. — British Patent 923653. 1963; GFR Patent 1134672. 1963.
113. Paterson, W. et al. — Canad.J.Chem., **39**(11):2324. 1961.
114. Fetter, N. and B.Bartócha. — Canad.J.Chem., **39**:2001. 1961.
115. Nazarova, L.M. — ZhOKh, **29**:2671. 1959.
116. Korneeva, G.K., D.Ya.Zhinkin, M.V.Sobolevskii, N.N. Korneev, E.I.Larikov, and N.V.Vladytskaya. — Soviet Patent 152573. 1962.
117. Zhinkin, D.Ya., G.K.Korneeva, N.N.Korneev, and M.V. Sobolevskii. — International Symposium on Organosilicon Chemistry, p. 311, Prague. 1965.
118. Zhinkin, D.Ya., G.K.Korneeva, N.N.Korneev, and M.V. Sobolevskii. — ZhOKh, **36**:350. 1966.

119. Stotskaya, L.L., N.N.Korneev, B.A.Krentsel', and D.Ya. Zhinkin. — Soviet Patent 168442. 1963.
120. Baranova, G.A., N.N.Korneev, B.A.Krentsel', and L.L. Stotskaya. — Vysokomolekulyarnye Soedineniya, **6**:1263. 1967.
121. Korneev, N.N., S.K.Goryunovich, and I.F.Leshcheva. — Plasticheskie Massy, No.10. 1968.
122. Zakharkin, L.I. and I.M.Khorlina. — Izvestiya AN SSSR, OKhN, **12**:2146. 1959.
123. Zakharkin, L.I. and O.Yu.Okhlobystin. — Izvestiya AN SSSR, OKhN, p.1006. 1958; Soviet Patent 110920. 1958.
124. Sander, M. — Angew.Chem., **73**:67. 1961.
125. Hock, H. and F.Ernst. — Chem.Ber., **92**:2716. 1959.
126. Sladkov, A.M. et al. — Trudy NIISS, p.115, Goskhimizdat. 1958.
127. Razuvaev, G.A. and A.I.Graevskii. — Izvestiya AN SSSR, OKhN, **9**:1555. 1962.
128. Razuvaev, G.A. and A.I.Graevskii. — ZhOKh, **32**(3):1006. 1962.
129. Razuvaev, G.A., E.V.Mitrofonov, and G.G.Petukhov. — ZhOKh, **31**:2340. 1961.
129a. Milovskaya, E.B. et al. — Izvestiya AN SSSR, OKhN, p.1093. 1967.
130. Jenkner, H. — GFR Patent 1031306. 1958.
131. Miller, A. et al. — J.Org.Chem., **24**:627. 1959.
132. Zakharkin, L.I. and I.M.Khorlina. — Izvestiya AN SSSR, OKhN, **12**:2255. 1959; **6**:1144. 1961.
133. Zakharkin, L.I. and V.V.Gavrilenko. — Izvestiya AN SSSR, OKhN, **12**:2245. 1960.
134. Zakharkin, L.I., L.M.Sorokina, and I.M.Khorlina. — ZhOKh, **31**:3311. 1961.
135. Zakharkin, L.I. and I.M.Khorlina. — Izvestiya AN SSSR, OKhN, **3**:538. 1962.
136. Pino, P. et al. — Chim.Ind., **44**(5):529. 1962.
137. Gilman, H. and J.Nelson. — J.Am.Chem.Soc., **61**:743. 1939.
138. Adkins, H. and C.Scanley. — J.Am.Chem.Soc., **73**:2854. 1951.
139. Zakharkin, L.I. and L.A.Savina. — Izvestiya AN SSSR, OKhN, **2**:378. 1961.
140. Kriner, W., A.MacDiarmid, and E.Evors. — J.Am.Chem. Soc., **80**:1546. 1958.
141. Imp.Chem.Ind., French Patent 1220843. 1960.
142. Podall, H. — J.Am.Chem.Soc., **80**:5573. 1958.
143. Gootes, H. et al. — US Patent 271546. 1955.
144. Pozamantir, A.G. and M.V.Genusov. — Soviet Patent 124440. 1959.
145. Zakharkin, L.I., G.S.Kolesnikov, S.L.Davydova, and V.V.Gavrilenko. — Izvestiya AN SSSR, OKhN, **2**:364. 1961.
146. Kolesnikov, G.S. et al. — Izvestiya AN SSSR, OKhN, **5**:841. 1962.
147. Sakharovskaya, G.B., N.N.Korneev, E.I.Larikov, A.F. Zhigach, and A.F.Popov. — ZhOKh, **34**:3435. 1964.
148. Sakharovskaya, G.B., N.N.Korneev, E.I.Larikov, A.F. Zhigach, and R.I.Fedotova. — Soviet Patent 170493. 1964.
149. Hoffman, E. and W.Fornau. — Angew.Chem., **73**(16):578. 1961.
150. Hollwachs, W. and A.Schafarik. — Ann.Chem., **106**:206. 1859.
151. Zhigach, A.F., I.S.Antonov, E.B.Kazakova, and R.S. Fraiman. — Soviet Patent 110579. 1957.

152. Zatham, K. — British Patent 800609. 1958.
153. Ziegler, K. — Angew. Chem., **64**:323. 1952.
154. Nobis, J. — US Patent 2960515. 1960.
155. Chini, P. et al. — Chim. Ind., **44**:1220. 1962.
156. Krause, E. and B. Wendt. — Ber., **56**:466. 1923.
157. Sladkov, A.M., I.Ya. Yavich, and P.K. Lunev. — Trudy NIISS, No. 1:104. 1958.
158. Bergwerkgesellschaft Hibernia, French Patent 1171028. 1959; Italian Patent 5741028. 1959; Italian Patent 574919. 1958; Austrian Patent 197833. 1958.
159. Reinheckel, H. — Tetrahedron Letters, No. 26:1939. 1964.
160. Bartocha, B.U. et al. — J. Am. Chem. Soc., **83**:2202. 1961.
161. Mangham, J. — US Patent 2969382. 1961.
162. Shapiro, H. and E. Dewitt. — US Patent 2987534. 1961.
163. Jenkner, H. — GFR Patent 944249. 1956.
164. Pino, P. et al. — Angew. Chem., **70**:599. 1958.
165. Pino, P. et al. — Ann. chim., **48**:1427. 1958.
166. Ramsden, H. — US Patent 3010895. 1961; British Patent 878130. 1961.
167. Bos, H. — GFR Patent 1018865. 1958; US Patent 2996529. 1961; British Patent 819380. 1959; Italian Patent 564082. 1957.
168. Zakharkin, L.I. and O. Yu. Okhlobystin. — ZhOKh, **31**(11):3662. 1961.
169. Yakubovich, A. Ya. and G.V. Motsarev. — Doklady AN SSSR, **88**:87. 1953.
170. Zakharkin, L.I. and L.A. Savina. — Izvestiya AN SSSR, OKhN, **5**:824. 1962.
171. Zakharkin, L.I. and L.A. Savina. — Izvestiya AN SSSR, OKhN, **6**:1039. 1960.
172. Zakharkin, L.I. — Izvestiya AN SSSR, OKhN, **2**:253. 1962.
173. Moretti, G. et al. — French Patent 1321239. 1963.
174. Goodrich-Gulf Chemicals, Inc., British Patent 878746. 1961.
175. Enfin, H. and A. Faillebin. — Comp. rend., **174**:112. 1922.
176. Hamprecht, G. and H. Muhlbauer. — US Patent 2867643. 1959; GFR Patent 1004179. 1957.
177. Badische Anilin u. Soda-Fabrik, British Patent 858498. 1959; GFR Patent 1070179. 1960.
178. Badische Anilin u. Soda-Fabrik, Italian Patent 560970. 1957; French Patent 1159087. 1958.
179. Ziegler, K. and H. Gellert. — US Patent 2826598. 1958; GDR Patent 13734. 1957; Dutch Patent 88075. 1958.
180. Köster, R. and W. Larbig. — GFR Patent 1124038. 1962.
181. Ziegler, K. et al. — British Patent 777701. 1957; GFR Patent 920071. 1957.
182. Ziegler, K. et al. — GFR Patent 918928. 1954.
183. Robinson, R. and E. Pritchett. — US Patent 2915542. 1959.
184. Wartik, T. and H. Schlesinger. — J. Am. Chem. Soc., **75**:835. 1953.
185. Ziegler, K. et al. — Angew. Chem., **67**:424. 1955.
186. Popov, A.F., N.N. Korneev, A.F. Zhigach, G.I. Volkov, and L.M. Antipin. — Soviet Patent 159838. 1962.

187. Topchiev, A.V., N.N.Korneev, A.F.Popov, and G.S. Shvinderman. — Soviet Patent 158678. 1962.
188. Korneev, N.N., G.S.Shvinderman, and L.I.Red'kina. — Vysokomolekulyarnye Soedineniya, **7**(9):1604. 1965.
189. Hall, F. and W.Nash. — Inst. Petr. Technology, **23**:679. 1937; **24**:471. 1938.
190. Ruthruff, R. — US Patent 2271956. 1942.
191. Robinson, J.et al. — French Patent 1291794. 1962.
192. Feighner, G. — French Patent 1322590. 1963.
193. Ziegler, K. — British Patent 916714. 1963.
194. Ziegler, K. and K.Zosel. — GFR Patent 1104509. 1961; French Patent 1273795. 1961; Austrian Patent 228799. 1963.
195. Nesmeyanov, A.N. et al. — Izvestiya AN SSSR, OKhN, **6**:1034. 1959.
196. Ethyl Corp., British Patent 835555. 1960.
197. Redman, H. — US Patent 2886581. 1959.
198. Sumitomo Chemical Co. Ltd., British Patent 834379. 1960.
199. Sumitomo Chemical Co. Ltd., US Patent 3016396. 1962.
200. Dobratz, E. — US Patent 3026345. 1962.
201. Redman, H. and B.Rouge. — US Patent 2787626. 1957.
202. Zakharkin, L.I., V.V.Gavrilenko, and O.Yu.Okhlobystin. — Izvestiya AN SSSR, OKhN, **1**:100. 1958.
203. Zhigach, A.F., A.F.Popov, L.D.Vishnevskii, L.M.Antipin, N.N.Korneev, and E.P.Bezukh. — Soviet Patent 125563. 1959.
204. Antipin, L.M., A.F.Zhigach, A.F.Popov, M.L.Rudkovskii, L.D.Vishnevskii, N.N.Korneev, and A.L.Kolpakov. — Soviet Patent 135486. 1960.
205. Zhigach, A.F., A.F.Popov, D.N.Sil'vestrov, M.A.Aronov, E.I.Larikov, L.M.Antipin, S.E.Nazarov, and N.N.Korneev — Soviet Patent 172780. 1962.
206. Korneev, N.N., A.F.Popov, and A.F.Zhigach. — Khimicheskaya Promyshlennost', **9**:645. 1962.
207. Popov, A.F., E.I.Larikov, L.M.Antipin, A.P.Fokin, and N.N.Korneev. — Obshchee Mashinostroenie, **11**(1). 1962.
208. Popov, A.F., L.M.Antipin, N.N.Korneev, A.F.Zhigach, and A.P.Fokin. — Vestnik Tekhniko-Ekonomicheskoi Informatsii NIITEKhIM, **6**:26. 1963.
209. Korneev, N.N. Author's Summary of Thesis. INKhS AN SSSR im. A.V. Topchieva. Moskva. 1963.
210. Antipin, L.M., L.D.Vishnevskii, A.F.Zhigach, and A.F. Popov. — Plasticheskie Massy, **1**:73. 1963.
211. Antipin, L.M. Author's Summary of Thesis. Moskovskii Neftyanoi Institut im. Gubkina. 1963.
212. Zhigach, A.F., A.F.Popov, L.D.Vishnevskii, and N.N. Korneev. — Khimicheskaya Promyshlennost', **4**:27. 1961.
213. Zhigach, A.F., A.F.Popov, L.D.Vishnevskii, N.N. Korneev, and E.P.Bezukh. — Soviet Patent 120173. 1958.
214. Popov, A.F., N.N.Korneev, and A.F.Zhigach. — Khimicheskaya Promyshlennost', **1**:25. 1966.
215. Zhigach, A.F., A.F.Popov, L.D.Vishnevskii, and L.M. Antipin. — Khimicheskaya Promyshlennost', **1**:24. 1962.
216. Asinger, F. et al. — Chem. Ber., **97**(9):2515. 1965.

217. Kryukov, S.I. et al. — Izvestiya Vuzov, Khimiya i Khimicheskaya Tekhnologiya, **1**:86. 1958.
218. Zhigach, A.F., I.S.Antonov, E.B.Kazakova, and R.S. Fraiman. — Khimicheskaya Promyshlennost', **2**:31. 1959.
219. Korneev, N.N., A.F.Popov, A.F.Zhigach, and G.I.Volkov. — Khimicheskaya Promyshlennost', **3**:18. 1963.
220. Zhigach, A.F., A.F.Popov, and E.P.Bezukh. — Byulleten' Tekhniko-Ekonomicheskoi Informatsii NIITEKhIM, **11**:39. 1962.
221. Zakharkin, L.I. and I.M.Khorlina. — ZhOKh, **30**:1926. 1960.
222. Roha, M. and L.Kreider. — Polymer Sci., **38**(133):51. 1959.
223. Korneev, N.N., L.M.Antipin, A.F.Popov, L.D.Vishnevskii, and A.F.Zhigach. — Vestnik Tekhniko-Ekonomicheskoi Informatsii, **7**:11. 1963.
224. Zhigach, A.F., A.F.Popov, N.N.Korneev, V.V. Gavrilenko, E.I.Larikov, and L.I.Zakharkin. — Soviet Patent 168691. 1962.
225. Korneev, N.N., A.F.Popov, A.F.Zhigach, and G.I.Volkov. — Plasticheskie Massy, **4**:29. 1965.

Chapter II

COMPOUNDS OF TRANSITION METALS

The subject of this chapter is the preparation and properties of compounds of the most important transition metals, which are used as components in catalyst complexes. We shall discuss, in the first place, the alkoxides and the halides of these metals, and then their acetylacetonates and certain organic derivatives. The derivatives of other transition metals are discussed more briefly, but adequate literature references on the subject are given.

1. HALIDES AND ALKOXIDES OF TRANSITION METALS

Most halides of transition metals (except for a few salts of titanium and vanadium) are crystalline compounds which form crystalline hydrates with water. Many of them fume in the air and decompose at elevated temperatures with evolution of the halogen and formation of salts of lower valencies. At elevated temperatures all metal halides react with water, with formation of hydroxides or of the corresponding metal anion. Compounds of this kind very typically form complexes with hydrohalic acids, their salts, bases, Lewis acids, etc.

Alkoxymetals and aryloxymetals are mostly colorless crystalline substances or viscous liquids which are stable on storage. Many of them are readily soluble in most organic solvents.

Titanium halides and alkoxides have been studied in most detail; halides and alkoxides of other metals have not been adequately studied.

Titanium chlorides

Titanium tetrachloride

Physical properties. Under normal conditions titanium tetrachloride is a monomolecular, mobile liquid, strongly fuming in the air. It forms white crystals when solid. Electron diffraction studies indicate that the $TiCl_4$ molecule is a regular tetrahedron, with the titanium atom at its center. The distances between the chlorine atoms are 3.61 Å, while the distance between titanium and chlorine atoms is 2.21 Å /1/. The volume of the molecule is $5.41 \cdot 10^{-24}$ cm^3. The dipole moment of the compound is zero /2/.

The IR spectrum of $TiCl_4$ was studied by Fowler /3/, while its UV spectrum was studied by Hukumoto /4/; its Raman spectrum has also been investigated /5/.

TABLE 3. Vapor pressure of titanium tetrachloride as a function of temperature

Temperature, °C	Vapor pressure, mm Hg	Temperature, °C	Vapor pressure, mm Hg	Temperature, °C	Vapor pressure, mm Hg
10	5.54	60	62.15	120	493.80
20	10.05	80	134.0	135.8	760
40	26.50	100	264.55		

Titanium tetrachloride melts at −23°C /6/ and boils at 135.8°C under atmospheric pressure /7/. The vapor pressures of the compound at different temperatures are shown in Table 3.

TABLE 4. Density of liquid $TiCl_4$ as a function of temperature

Temperature, °C	Density, g/ml	Temperature, °C	Density, g/ml	Temperature, °C	Density, g/ml
-18.1	1.79167	20	1.7277	79.77	1.62517
0	1.76087	39.97	1.69371	100.0	1.58914
10.5	1.744	59.86	1.65974	136	1.522

Titanium tetrachloride does not decompose on being boiled, and its vapors are stable even at temperatures above its normal boiling point; some chlorine is evolved at about 2000°C /8/.

The energy of dissociation of the molecule, calculated from the long-wave UV absorption boundary for the probable reaction

$$TiCl_4 + h\nu \longrightarrow TiCl_3 + Cl^*$$

is 87 kcal for gaseous $TiCl_4$ /9/. The densities of liquid $TiCl_4$ are shown in Table 4. The volume expansion between −18 and 110°C may be expressed by the formula

$$V_t = V_0\,(1 + 9.6457\cdot 10^{-4}t + 6.026\cdot 10^{-7}t^2 + 5.94\cdot 10^{-9}t^3)$$

TABLE 5. Viscosity of liquid $TiCl_4$ as a function of temperature

Temperature, °C	Viscosity, poise	Temperature, °C	Viscosity, poise	Temperature, °C	Viscosity, poise
-15	0.0126	20	0.00826	40	0.00702
0	0.01012	30	0.00756	50	0.00645
10	0.00912				

Table 5 shows the viscosity variation of the compound with temperature /10/; other physical parameters are listed in Table 6.

TABLE 6. Physical constants of titanium tetrachloride

Parameter	Value	References
Latent heat of fusion, cal/g	11.75	11
Latent heat of vaporization, cal/g		
at 25°C	47.22	7
at 135.8°C	45.43	
Average heat capacity at 13 — 99° C, cal/g . . .	0.18812	12
Refractive index, n_D^{16}	1.6059	13
Molar refraction MR_D	65.23	14
Critical temperature, °C	358.0	14
Entropy of liquid $TiCl_4$, $S_{298.1\,°K}$	59.51	15

Main chemical properties. Hydrogen reacts with titanium tetrachloride at 500 — 800°C with formation of the violet-colored $TiCl_3$ /16/:

$$2TiCl_4 + H_2 \rightleftarrows 2TiCl_3 + 2HCl$$

Above 900°C, in the presence of excess hydrogen, the reaction proceeds to the stage of bivalent and even monovalent titanium /17, 18/. At high hydrogen pressures titanium hydride is formed as a black powder /19/.

Metals of Groups I — III react with $TiCl_4$ to reduce Ti(IV) to the metallic state. Thus, for instance, sodium reduces $TiCl_4$ to $TiCl_2$ in the cold:

$$TiCl_4 + 2Na \longrightarrow TiCl_2 + 2NaCl$$

At 130 — 150°C titanium tetrachloride is reduced to titanium /20/. Aluminum in the presence of $AlCl_3$ reduces $TiCl_4$ to $TiCl_3$ /21/:

$$3TiCl_4 + Al \longrightarrow 3TiCl_3 + AlCl_3$$

Nonmetals react with $TiCl_4$ only with much difficulty. The tetrachloride does not react with carbon /21/; nitrogen reacts only after activation by silent electric discharge, with formation of titanium chloronitride TiNCl /22/.

Phosphorus, arsenic and antimony reduce $TiCl_4$ to $TiCl_3$ /21, 23/:

$$3TiCl_4 + P \longrightarrow 3TiCl_3 + PCl_3$$

When a mixture of $TiCl_4$ vapors and oxygen was passed through a glowing tube at above 550°C, TiO_2 and $Ti_2O_3Cl_2$ were obtained /24/:

$$TiCl_4 + O_2 \longrightarrow TiO_2 + 2Cl_2$$

$$4TiCl_4 + 3O_2 \longrightarrow 2Ti_2O_3Cl_2 + 6Cl_2$$

It was noted that $TiCl_4$ does not react with oxygen below 550°C. It reacts with sulfur in the presence of aluminum trichloride, with formation of $TiCl_3$ and sulfur monochloride:

$$2TiCl_4 + S_2 \longrightarrow 2TiCl_3 + S_2Cl_2$$

Except for fluorine, halogens are miscible with titanium tetrachloride in all proportions. The action of fluorine produces an exchange of halogen atoms with the evolution of elementary chlorine:

$$TiCl_4 + 2F_2 \longrightarrow TiF_4 + 2Cl_2$$

When $TiCl_4$ is introduced into water, a large amount of heat is evolved, and the voluminous titanium hydroxide precipitates out. It was found that when 1 mole of $TiCl_4$ is added to 1600 moles of water at 17°C, 57,870 calories are evolved /24/. At low temperatures the main product is the pentahydrate $TiCl_4 \cdot 5H_2O$. This product is orange-yellow and is strongly hygroscopic. It decomposes in water to form an orange-yellow, viscous liquid. The pentahydrate $TiCl_4 \cdot 5H_2O$ may be preserved only in the cold and in a concentrated solution. It was shown by Luchinskii /24/ that the introduction of hydrochloric acid slows down the hydrolysis process considerably. It was shown /25/ that this hydrate may exist for a while in the nonhydrolyzed form in 0.5 N hydrochloric acid solution. On further dilution hydrolysis takes place immediately, via the following successive steps:

$$TiCl_4 \cdot 5H_2O \longrightarrow TiCl_3(OH) \cdot 4H_2O + HCl$$
$$TiCl_3(OH) \cdot 4H_2O \longrightarrow TiCl_2(OH)_2 \cdot 3H_2O + HCl \text{ etc.}$$

The intermediate products may be isolated in the cold. Thus, for instance, the isolated $TiCl_3(OH) \cdot 4H_2O$ and $TiCl_2(OH)_2 \cdot 3H_2O$ are yellow substances which are unstable in water; $TiCl(OH)_3 \cdot 2H_2O$ is a white, hygroscopic substance which does not react with water. Nevertheless, in the presence of a large excess of water at 0°C a gel-like mass of the composition $Ti(OH)_4 \cdot H_2O$ separates out /26/. Titanium tetrachloride is rapidly hydrolyzed by boiling water to metatitanic acid:

$$TiCl_4 + 3H_2O \longrightarrow H_2TiO_3 + 4HCl$$

Gaseous hydrogen sulfide reduces $TiCl_4$ to $TiCl_3$ /27/:

$$2TiCl_4 + H_2S \longrightarrow 2TiCl_3 + 2HCl + S$$

If $TiCl_4$ is mixed with liquid H_2S, the brown-colored thiohydrate $2TiCl_4 \cdot H_2S$, soluble in excess H_2S, precipitates out first. At low temperatures the compounds $TiCl_4 \rightarrow H_2S$ and $TiCl_2 \cdot 2H_2S$ are formed; these are yellow crystals. All thiohydrates of tetravalent titanium can be distilled. At elevated temperatures they react with excess hydrogen sulfide, forming products of thiohydrolysis. The final product of thiohydrolysis is thiotitanic acid /28/:

$$TiCl_3(SH) + 3H_2S \longrightarrow Ti(SH)_4 + 3HCl$$

When hydrogen sulfide is reacted with boiling $TiCl_4$, titanium thiochloride and titanium sulfide are formed:

$$TiCl_4 + 2H_2S \longrightarrow TiS_2 + 4HCl$$

Titanium tetrachloride reacts with ammonia to form complexes $TiCl_4 \cdot 4NH_3$ or $TiCl_4 \cdot 6NH_3$ /29, 30/. The ammoniate $TiCl_4 \cdot 4NH_3$ is a very hygroscopic red-brown powder, which decomposes on being heated, with the evolution, first, of ammonia and ammonium chloride and then of hydrogen chloride. The next stage is the vaporization of $(NH_4)_2TiCl_6$. When gaseous ammonia is reacted with $TiCl_4$ vapors in an atmosphere of hydrogen, $TiCl_4 \cdot 6NH_3$ separates out as a very hygroscopic, bulky, dark yellow powder. Titanium tetrachloride readily dissolves gaseous hydrogen chloride. It has been pointed out /31/ that two molecular compounds are formed: $TiCl_4 \cdot 2HCl$ (m.p. $-30.8°C$) and $TiCl_4 \cdot HCl$ (m.p. $-86°C$). Other hydrogen halides react with $TiCl_4$ to form the corresponding titanium halide:

$$TiCl_4 + 4HX \longrightarrow TiX_4 + 4HCl$$

Phosphorus, arsenic, antimony, silicon, carbon and tin halides are miscible with $TiCl_4$ in all proportions. Sulfur halides form three compounds with $TiCl_4$: $2TiCl_4 \cdot SCl_4$ /32/, $TiCl_4 \cdot SCl_4$ /33/ and $TiCl_4 \cdot 2SCl_4$ /34/.

Reactions with organic compounds. Titanium tetrachloride is soluble in saturated liquid hydrocarbons. If the hydrocarbon is pure, the solution remains colorless and transparent for a long time. Solutions of $TiCl_4$ in unpurified paraffinic hydrocarbons become turbid and assume a light yellow coloration, which then becomes dark yellow. Titanium tetrachloride usually forms yellow-colored solutions in benzene and toluene; it is also soluble in halogenated aliphatic and aromatic hydrocarbons.

Titanium tetrachloride reacts with alcohols with formation of titanium alcoholates. It was noted /35/ that the reaction between titanium tetrachloride and alcohols taken in excess, when conducted at an elevated temperature, usually results in the formation of dialkoxytitanium dichloride. Reaction between $TiCl_4$ and hydroxylated aromatic compounds results in a product, in which a number of hydroxyls have been replaced by chlorine; the exact number of such chlorine atoms depends on the structure of the initial product.

Titanium tetrachloride reacts with ethers to form complexes of the type $TiCl_4 \cdot (CH_3)_2O$. Prolonged action of light on solutions of $TiCl_4$ results in a reduction of titanium /36/. Titanium tetrachloride vigorously reacts with aldehydes and ketones, with formation of the corresponding addition products /37/. Reactions with a diketone such as acetylacetone yield crystalline needles of the composition $TiCl_4 \cdot CH_3COCH_2COCH_3$. In ethereal solutions chlorine atoms are successively substituted by diketone residues and HCl is evolved:

$$TiCl_4 + CH_3COCH_2COCH_3 \longrightarrow TiCl_3(CH_3COCHCOCH_3) + HCl \text{ etc.}$$

Titanium tetrachloride forms complexes by reactions with amines. Thus, for example, the aniline complex $TiCl_4 \cdot 4C_6H_5NH_2$ /38/ has been isolated. When $TiCl_4$ is reacted with free diphenylamine, greenish-brown crystals are obtained; dilute solutions of diphenylamine yield a greenish-yellow precipitate of the composition $TiCl_4 \cdot (C_6H_5)_2NH$ /38/.

Indole reacts with $TiCl_4$ to form the molecular compound $TiCl_4 \cdot 2C_8H_7N$ /39/. A benzene solution of $TiCl_4$ reacts with pyridine to form the yellow precipitate of $TiCl_4 \cdot 2C_5H_5N$ /38/; in ether the unstable complex $TiCl_4 \cdot 6C_5H_5N$ separates out as a brown-colored, amorphous mass /40/. With quinoline the brown $TiCl_4 \cdot 2C_9H_7N$ is obtained.

Titanium tetrachloride reacts with organometallic compounds to form alkyl derivatives, which are unstable and decompose above $-50°C$, with evolution of $TiCl_3$. Titanium tetrachloride may be alkylated by alkyls of alkali metals or by alkyls of the metals of Groups II — IV /41 — 43/:

$$MeR + TiCl_4 \longrightarrow MeCl + RTiCl_3$$

Reactions of metal alkyls with titanium tetrachloride will also be discussed in Chapter III.

Preparation. As is well known, the general method for the preparation of titanium tetrachloride is the chlorination of titanium-containing slags, mostly with subsequent condensation of the titanium chloride vapors formed.

Studies of the chlorination of titanium oxides, titanium carbide /210/, and titanium slags containing oxides of other metals such as calcium and magnesium oxides /211/ showed that the chlorination is a complex process which includes two main stages. Initially, lower titanium oxides are chlorinated, with formation of titanium tetrachloride and titanium dioxide:

$$2TiO + 2Cl_2 \longrightarrow TiCl_4 + TiO_2$$

This is followed by the chlorination of TiO_2.

The reaction between chlorine and titanium dioxide takes place in a thin sorption layer, which is formed around the carbon and TiO_2 particles; it was suggested by Reznichenko and Solomakha /212/ that the chlorinating agent in the layer is a sorptional complex of the type of $COCl_2$, which reacts with TiO_2 to form $TiCl_4$ and CO_2. Depending on the flow rate of the gas, carbon dioxide may be desorbed together with $TiCl_4$. The impurities in the titanium slag, say magnesium or calcium oxides, play an activating part in the chlorination of titanium oxides. The mechanism of this effect is not yet clear, but may consist in the acceleration of the desorption of $TiCl_4$.

It was found /212/ in a study of the chlorination of calcium metatitanate that calcium oxide is chlorinated first. Below 725°C, CO_2 is mainly evolved, while above 725°C the evolution of CO predominates:

$$CaO \cdot TiO_2 + Cl_2 + {}^1/_2C \xrightarrow{< 725°C} CaCl_2 + {}^1/_2CO_2 + TiO_2$$

$$CaO \cdot TiO_2 + Cl_2 + C \xrightarrow{> 725°C} CaCl_2 + CO + TiO_2$$

It was shown that in the chlorination of the ore concentrate, the main monomineral phase of which is calcium metatitanate (perovskite), 74% of TiO_2 and 96.6% of CaO were chlorinated at 700°C.

Depending on the nature of the titanium slag being chlorinated, the titanium tetrachloride product may contain various amounts of impurities; these are chlorides of other metals, such as iron, aluminum, magnesium, manganese, calcium and silicon. Under industrial conditions all these chlorides are condensed by sprinkling with cooled titanium tetrachloride. The resulting condensate is a slurry, which contains suspended fine particles of solid $AlCl_3$, $FeCl_3$, etc. Some of the solid chlorides are dissolved in titanium tetrachloride. It was reported by Galitskii and Shadskii /213/ that a technical grade titanium tetrachloride contained the following percentage amount of impurities:

$SiCl_4$	up to 2
$TiOCl_2$	0.01 — 0.05
$VOCl_3$	0.05 — 0.2
HCl	0.01 — 0.2
$COCl_2$	0.01 — 0.09
$MgCl_2$	0.03 — 0.1
$MnCl_2$	0.02 — 0.07

It was also found that the solubility of aluminum chloride, iron chlorides, etc. in $TiCl_4$ increases with temperature.

The construction of the equipment for the industrial production of titanium tetrachloride involves certain difficulties which are due to the specific properties of this compound. In addition, since the slurry contains solid chlorides of extraneous metals and liquid $SiCl_4$ suspended in titanium tetrachloride, silicon tetrachloride must be separated from the solid matter by settling, centrifugation, filtration or rectification. The removal from titanium tetrachloride of impurities such as vanadium chlorides or residual aluminum chloride involves the use of physicochemical techniques of purification by formation of complex compounds; this is achieved by introducing copper powder, moist activated charcoal, etc. into the solution, with subsequent settling and filtration of the solid matter.

The most typical property of liquid $TiCl_4$ and $TiCl_4$-based slurries is the tendency of $TiCl_4$ (which has a very high vapor pressure at normal temperatures) to react with atmospheric moisture to give hydrogen chloride and solid titanium oxychlorides, which interfere with the normal transport of titanium tetrachloride through the liquid and gas ducts. The penetration of titanium tetrachloride into the atmosphere has an adverse effect on the working hygiene; accordingly, the equipment used in the process must be airtight.

A second, not less important property of titanium tetrachloride is that it corrodes the steels and linings commonly employed in equipment construction. Accordingly, the equipment used in the manufacture of titanium tetrachloride must be made of special brands of steel.

Titanium trichloride

Titanium trichloride is a crystalline powder, which is usually violet or brown-colored. It is assumed that the true color of $TiCl_3$ is black, and the violet coloration is produced by traces of moisture /44/.

Titanium trichloride may exist in four crystalline modifications, which are known at present as α-, β-, γ- and δ-forms /45/. Figure 1 shows all these four modifications. The three crystalline modifications of violet $TiCl_3$ (α-, γ- and δ-forms) are built of alternate titanium layers, each of which is sandwiched between two layers of chlorine atoms. These layers may be differently disposed relative to one another, which accounts for the differences between the three modifications. The brown β-modification has no layered structure and may be represented as a linear polymer. It was shown by Korotkov and Li Tsung-ch'ang /47/ that the β-$TiCl_3$ structure may be converted into the α-form by external forces. These workers showed by electron-microscopic investigations that when β-$TiCl_3$ particles are heated, the particle size increases from $1-3\,\mu$ to $3-10\,\mu$. Prior to heating β-$TiCl_3$

forms fairly big needle clusters; when held at 200°C for 20 hours, some of the material recrystallizes to form hexagonal plates (α-$TiCl_3$). Heating at 300 and 400°C yields larger-sized crystals in the form of hexagonal prisms. When the temperature is further increased, $TiCl_3$ disproportionates to form $TiCl_2$ and volatile $TiCl_4$. The conversion of β-$TiCl_3$ into α-$TiCl_3$, which was revealed by electron microscope studies, was then confirmed by X-ray analysis, while the change in the particle size was confirmed by measuring the specific surfaces of various $TiCl_3$ samples by the nitrogen adsorption method.

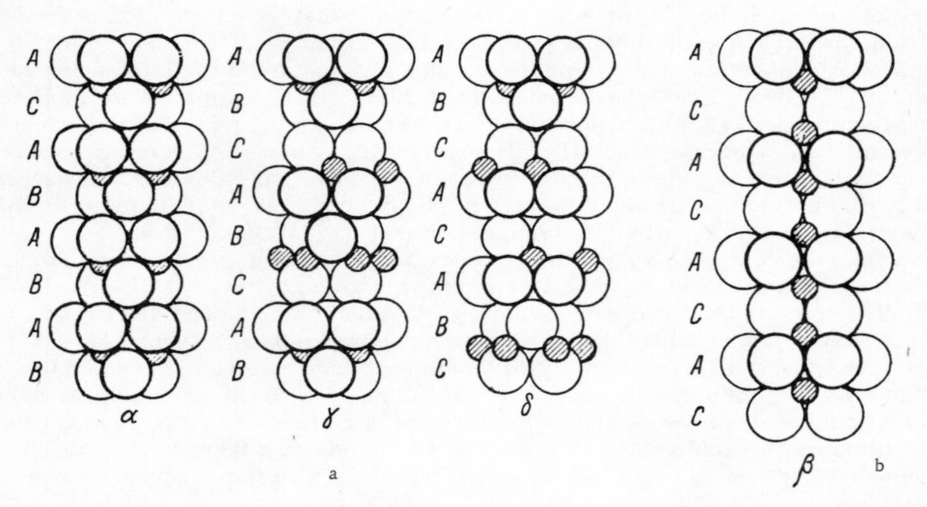

FIGURE 1. Structure of the crystalline modifications of $TiCl_3$:

a — violet $TiCl_3$; b — brown $TiCl_3$.

Titanium trichloride is paramagnetic; its density (at an unspecified temperature) is 2.656 /48/. At low temperatures $TiCl_3$ becomes antiferromagnetic. It distils without decomposition at pressures below 1 mm Hg and separates out as dark violet prisms on the cool parts of the apparatus /21/. Under pressures of the order of 10^{-3} mm Hg, when the temperature is slowly raised to 440°C, the crystals gradually sublime and condense again as thin, violet-colored platelets, 3 — 4 mm in diameter /44/. Dijkgraaf and Rousseau /49/ studied the absorption spectra of α-$TiCl_3$ crystals in polarized light. Titanium trichloride does not show any changes when heated under atmospheric pressure for several hours at 500°C /21/. Titanium trichloride can be distilled in a mixture of hydrogen and $TiCl_4$ vapors at 600°C /50/.

Titanium trichloride is readily soluble in water and in alcohol, with formation of violet or green-colored solutions; it is insoluble in ether. It is sparingly soluble in hydrochloric acid /51/. It is insoluble in $TiCl_4$, $CHCl_3$, CCl_4, CS_2 and benzene /52/. The color of aqueous solutions of $TiCl_3$ largely depends on its concentration of hydrochloric acid. Thus, when a solution of $TiCl_3$ is strongly diluted with water, the violet color disappears. However, when concentrated hydrochloric acid is added to the solution in order to suppress the dissociation, the violet coloration reappears /53/. If excess water is evaporated from a solution of $TiCl_3$ or if the solution is saturated with HCl, the violet hexahydrate $TiCl_3 \cdot 6H_2O$ can be obtained. There is also a green form of $TiCl_3$ hydrate, but it is very unstable /54/.

On passing gaseous ammonia through $TiCl_3$, a white powder of the ammoniate $TiCl_3 \cdot 6NH_3$ separates out. On being heated to 300°C, it is converted to the very reactive $TiCl_3 \cdot 2NH_3$ /55/.

Preparation of α-$TiCl_3$. Until recently, titanium trichloride was not properly studied, but its use as the main component of stereospecific catalyst complexes resulted in the search for new ways of preparation of $TiCl_3$, and development of industrial syntheses based on reactions already known.

The main starting material in the preparation of $TiCl_3$ is titanium tetrachloride. The other reactant is any reducing agent such as hydrogen, sodium, magnesium, aluminum, silicon, titanium, etc.

One of the most interesting methods of preparation of titanium trichloride is the reaction between titanium tetrachloride and hydrogen, which has found industrial application outside the Soviet Union /214/. Titanium trichloride was prepared in this way at 1000 — 1200°C:

$$2TiCl_4 + H_2 \rightleftarrows 2TiCl_3 + 2HCl$$

The hydrogen chloride formed during the reaction must be continuously removed in order to shift the reaction equilibrium so as to favor the formation of the violet $TiCl_3$. The equilibrium of the reaction was the subject of a detailed study /215/. Metallic titanium was put in the path of the gaseous reaction products in order to absorb HCl, as a result of which the yield of $TiCl_3$ increased considerably /216/. A continuous process for the preparation of $TiCl_3$ has now been developed: hydrogen is made to react with $TiCl_4$ in an electric arc discharge. The product is collected in a special vessel and the issuing hydrogen chloride is passed over heated titanium. The titanium tetrachloride and hydrogen which are formed as a result are recycled /217/.

The hydrogen method for the preparation of violet $TiCl_3$, though used on an industrial scale /218/, suffers from a number of serious drawbacks, including the need for working at an elevated (1000 — 1200°C) temperature, a thorough purification of the hydrogen employed, utilization of a large excess of one of the components of the reaction, the need for special materials, etc. A search was accordingly instituted for other technologically suitable methods of preparation of titanium trichloride.

Magnesium and zinc /212/ were used as reductants of $TiCl_4$. When this was done, $TiCl_3$ was formed at 200 — 400°C and the conversion of the reducing metal was low. Similar results were obtained when mercury was used as reducing agent.

Titanium tetrachloride is reduced to the trichloride at an intensive rate by sodium or aluminum. The reduction by aluminum is conducted at 200°C in the presence of a small amount of added $AlCl_3$, when a $TiCl_3$ — $AlCl_3$ complex is formed /220/. Titanium tetrachloride is reduced by metallic sodium to titanium trichloride, but the product contains up to 50% sodium chloride /221/. This method is suitable for the manufacture of titanium trichloride on an industrial scale, and is attractive by its simplicity /222/. Thermal reduction of titanium tetrachloride by sodium is performed in two equipment units. The unit which accommodates metallic sodium is made airtight, is inserted in an electric furnace and heated to 500 — 550°C, and

$TiCl_4$ is then passed over the molten sodium. The rate of the process is adjusted by varying the temperature in the reactor, the feed rate of $TiCl_4$ and the reactor pressure. The resulting salt mixture is pressed into an airtight mold; the melt in the mold is allowed to cool, the mold is opened and the resulting bar of $TiCl_3$ is ground.

Interesting studies were carried out /223/ on the preparation of titanium trichloride by reducing titanium tetrachloride by $TiO_2 - C$ or $TiO - C$ mixtures:

$$3TiCl_4 + TiO_2 + 2C \rightleftarrows 4TiCl_3 + 2CO$$
$$3TiCl_4 + TiO + C \rightleftarrows 4TiCl_3 + CO$$

These reactions are best performed at 1000 — 1500°C. The reaction products form a gaseous mixture consisting of $TiCl_3$, $TiCl_4$ and CO. The following side reaction may also take place:

$$8TiCl_3 + CO \rightleftarrows 6TiCl_4 + TiO + TiC$$

As the temperature decreases, this equilibrium is shifted to the right. On slow cooling the gaseous compounds which contain $TiCl_3$ and CO may be converted into other products.

A patent /224/ has been taken out for the preparation of $TiCl_3$ by the reaction of TiC with $TiCl_4$:

$$3TiCl_4 + TiC \rightleftarrows 4TiCl_3 + C$$

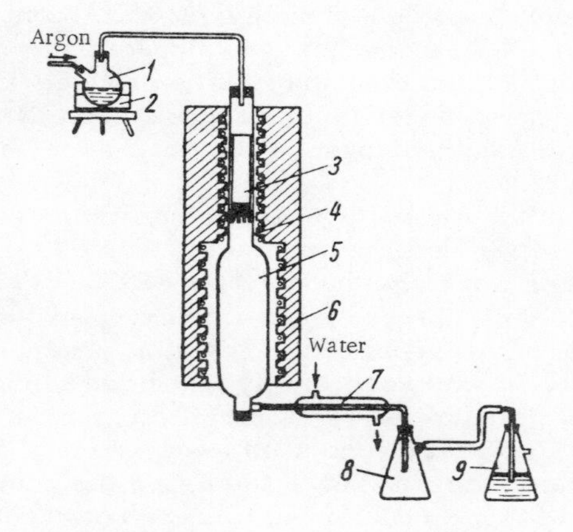

FIGURE 2. Laboratory apparatus for the preparation of titanium trichloride:

1 — evaporator of titanium tetrachloride; 2 — glycerin bath; 3 — graphite beaker with silicon; 4 — reactor; 5 — condenser; 6 — electric furnace; 7 — heat exchanger; 8 — collector of $TiCl_4$ and $SiCl_4$; 9 — absorber bottle for volatile chlorides.

titanium oxides and carbon; it utilizes the fact that the oxides react with the carbon above 900—1000°C, when the carbide and carbon monoxide are produced /225/.

One of the most promising methods for the preparation of titanium trichloride is the reaction between $TiCl_4$ and Ti:

$$3TiCl_4 + Ti \longrightarrow 4TiCl_3$$

This reaction was checked by Ryabov and Zviadadze and was found suitable for preparative purposes /226/. The reaction begins above 500°C. According to these workers, the first stage in the process is the reaction between Ti and $TiCl_4$, with formation of $TiCl_2$:

$$TiCl_4 + Ti \longrightarrow 2TiCl_2$$

after which $TiCl_2$ reacts with $TiCl_4$ to form $TiCl_3$.

Seelbach /227/ subsequently patented a method for the preparation of $TiCl_3$ by reduction of titanium tetrachloride with metallic titanium. In order to prepare titanium trichloride in high yields, metallic titanium should be present in excess. The reaction proceeds in a satisfactory manner at 400 — 500°C under a pressure of above 100 atm.

Furman and Lavrova /228/ developed an interesting process for the preparation of the violet $TiCl_3$ from titanium tetrachloride and silicon. These workers showed that the sole product of the reaction is $TiCl_3$ which contains a small amount of $TiCl_2$. A scheme for a laboratory scale preparation of $TiCl_3$ is shown in Figure 2. The optimum reaction temperature is 900°C. The conversion of $TiCl_4$ is low: only 15 — 17% per pass. The product is technical grade α-$TiCl_3$, 98 — 99% pure, which contains less than 0.5% $TiCl_4$ and less than 1.0% insoluble admixtures /214/.

Preparation of β-$TiCl_3$. One of the simplest methods for the preparation of the brown variety (β-form) of titanium trichloride is the reaction between an alkylaluminum compound and a solution of $TiCl_4$ in a hydrocarbon.

Korotkov and Li Tsung-ch'ang /229/ prepared β-$TiCl_3$ by the reaction between $Al(C_2H_5)_2Cl$ and $TiCl_4$, taken in equimolecular amounts. The resulting precipitate of β-$TiCl_3$ was filtered at room temperature onto a No. 4 sintered glass filter and washed several times with the anhydrous solvent. The β-$TiCl_3$ thus obtained contained small amounts of $TiCl_4$, aluminum and organic compounds. The organic compounds should be removed by petroleum ether extraction conducted at 80°C.

High purity β-$TiCl_3$ preparations may be obtained by decomposition of methyltrichlorotitanium /230/.

Titanium dichloride

Titanium dichloride is a black-colored, crystalline compound, with a hexagonal lattice /56/. The compound is unstable in the presence of $TiCl_4$ even at room temperature /57/; it burns in the air with the evolution of $TiCl_4$ vapors and formation of TiO_2 /58/. In vacuum, even under very low residual pressures, titanium dichloride is not markedly volatile at temperatures above 600°C /21/. Titanium dichloride vigorously reacts with water, with evolution of hydrogen and formation of a yellow solution, the properties of which resemble those of the partly oxidized solution of $TiCl_3$. When the amount of water is very small, the compound is prone to spontaneous combustion /58/.

Titanium dichloride bursts into flame on being reacted with bromine, when it forms a liquid with b.p. 177°C. It would appear that the liquid consists of an azeotropic mixture of $TiCl_4$ and $TiBr_4$ /58/. Concentrated nitric acid oxidizes $TiCl_2$, and the reaction products are spontaneously flammable. In concentrated hydrochloric and sulfuric acids $TiCl_2$ dissolves to form a green-colored solution /57/. If $TiCl_4$ is added to this solution, it becomes violet-colored, indicating that bivalent titanium is present /55/.

Titanium dichloride is insoluble in ether, CS_2 and chloroform. It reacts with alcohol with evolution of hydrogen and formation of a yellow solution, which turns blue-black when treated with ammonia /58/.

Titanium dichloride reacts with gaseous ammonia with evolution of heat and formation of the pearl-gray $TiCl_2 \cdot 4NH_3$. A similar reaction takes place with ammonia at −78°C. The compound $TiCl_2 \cdot 4NH_3$ decomposes in vacuo at 300°C to give ammonia and ammonium chloride. The residue apparently contains titanium nitride /59/.

The specific magnetic susceptibility at −180, −78 and +15.6°C is $6.65 \cdot 10^{-6}$, $4.90 \cdot 10^{-6}$ and $4.30 \cdot 10^{-6}$ respectively /60/.

Other halides of titanium and of titanium group metals

Titanium halides such as $TiBr_4$, TiI_4 or $TiBr_3$ resemble in their properties the corresponding chlorides of titanium. Their hydrolysis mainly takes place according to the following scheme:

$$TiX_4 + 2H_2O \rightleftarrows Ti(OH)_4 + 4HX$$

where X is Br or I.

Another very typical property of titanium halides is their strong tendency to participate in addition reactions. Some of the physical properties of titanium halides $TiHal_4$ and $TiHal_3$ are shown in Table 7.

TABLE 7. Some physicochemical properties of titanium halides

Compound	Halogen	Color	M.p., °C	B.p., °C
TiX_4	F	Colorless	—	284
	Cl	"	−23	136
	Br	Yellow	39	230
	I	Dark red	150	377
TiX_3	Cl	Violet	Sublimes at 425–480°C	
	Br	Bluish-black	Decomposes at 400°C	
	I	Violet-black	Decomposes above 350°C	

Zirconium and hafnium halides are colorless crystalline compounds. Their hydrolysis may be described by the equation:

$$ZrX_4 + H_2O \rightleftarrows ZrOX_2 + 2HX$$

where X is Cl, Br or I.

Fluorides, unlike the other halides, react with water to form complex acids of the type $H_2[MOF_4]$ and for this reason do not, practically speaking, undergo hydrolysis, even on being heated. Zirconium and hafnium halides typically undergo sublimation on being heated. The vapor pressures of $ZrCl_4$, $ZrBr_4$, ZrI_4, $HfCl_4$ and $HfBr_4$ attain 760 mm Hg at 331, 357, 431, 317 and 322°C respectively. Their melting points can be determined only under elevated pressures and are, respectively, 437, 450, 499, 432 and 420°C /62/. Spiridonov et al. recently determined the crystalline structure of $ZrCl_4$ /61/. This salt crystallizes as tetrahedra; it has a molecular cubic lattice with $a = 10.32$ Å, coordination number 8, and Zr — Cl distance 2.32 ± 0.02 Å. The magnetic susceptibility is $0.301 \cdot 10^{-6}$ CGSM; the density at 15°C is 2.803 g/cm^3; the heat of formation of $ZrCl_4$ from the elements is 231.9 ± 0.51 Kcal/mole. Zirconium tetrachloride is insoluble in CCl_4, heptane and other similar solvents. If dissolved in benzene, formaldehyde, acetone or similar solvents, profound chemical transformations take place. Zirconium tetrachloride is also soluble in fused alkali and alkaline earth metal chlorides.

Zirconium tetrabromide crystallizes in a primitive cubic lattice with $a = 10.95 \pm 0.01$ Å /63/. It is much less stable to heat than $ZrCl_4$, and decomposition is already significant at 1245°C /64/, while the tetraiodide begins to decompose at 1000°C. Complete decomposition of $ZrBr_4$ occurs above 1450°C /64/.

Trivalent zirconium compounds are known. They are brown or black-colored modifications, which are formed when the corresponding ZrX_4 is heated with metallic aluminum in a sealed tube. They are all very strong reducing agents.

The blue-black $HfBr_3$ greatly resembles $ZrBr_3$ in its properties. It decomposes on being heated above 300°C in vacuo:

$$2HfBr_3 \longrightarrow HfBr_4 + HfBr_2$$

Halides of metals of Group V and Group VI

Halides of metals of the vanadium sub-group /62/

The most interesting compounds of Group V metal halides are vanadium halides. These compounds, like titanium halides, react very vigorously with water and with other hydroxylated compounds. Of the pentahalides, the fluoride only is known; it may be prepared by reacting the elements at 300°C and is a crystalline, colorless substance with a sublimation point of 111°C. Vanadium pentafluoride is fully hydrolyzed by water. Pentavalent vanadium also forms oxyhalides of the general formula $VOHal_3$. Of these VOF_3 forms yellowish-white crystals (m.p. 300°C, b.p. 480°C). With KF and NaF the complexes $3MF \cdot 2VOF_3$ are formed. The yellow $VOCl_3$, m.p. −77°C, b.p. 127°C, is readily hydrolyzed to vanadic acid. It may be prepared by the reaction between dry HCl and heated V_2O_5, in the presence of phosphorus pentoxide to bind the water formed. As distinct from VOF_3, vanadium oxytrichloride does not form complexes with alkali metal halides. Vanadium oxybromides are unstable and decompose on being heated. The orthobaric densities and critical parameters of $VOCl_3$ have also been determined /57/.

The oxyhalides formed by vanadium dioxide with the corresponding acids are very hygroscopic. Vanadyl chloride $VOCl_2$ is green when solid. It readily dissolves in water, forming a blue or brown-colored solution, depending on the experimental conditions. The brownish-black $VOBr_2$ forms blue solutions in water.

Vanadium tetrachloride, which is best prepared by direct heating of the elements, is a heavy red-brown liquid (m.p. $-26°C$, b.p. 152°C). It forms dimers when dissolved in CCl_4. On being heated, vanadium tetrachloride slowly decomposes into vanadium trichloride and chlorine; it reacts with water to form vanadyl chloride. It reacts with KCl, RbCl and CsCl at 400°C to form addition compounds, which are brown, red-pink and violet-colored respectively.

Vanadium tetrafluoride, which is prepared by prolonged heating of VCl_4 with anhydrous hydrofluoric acid, is a brown-colored powder. When heated above 300°C in a nitrogen atmosphere, it decomposes into VF_3 and VF_5. Vanadium tetrafluoride is hygroscopic and is easily hydrolyzed with the formation of VOF_2.

Vanadium trichloride is obtained by thermal decomposition of VCl_4. It forms red-violet nonvolatile crystals, which are readily soluble in water, in which they form green-colored solutions.

In the same manner vanadium tribromide and triiodide can also be prepared. They resemble vanadium trichloride in their properties, but are less stable.

Vanadium dichloride, which is prepared by passing a mixture of hydrogen and VCl_4 vapors through a red-hot tube, forms pale green crystals. It dissolves in water, forming a violet-colored solution, which then again becomes green owing to the oxidation of V(II) to V(III). Vanadium dichloride forms blue solutions in alcohol and green-yellow solutions in ether. The brown-colored VBr_2 and the pink-colored VI_2 display similar properties.

Tantalum and niobium halides do not differ in many respects from the vanadium compounds. Both they and vanadium salts may form oxyhalides, in which the metal may be bivalent or trivalent. However, metal pentahalides are the most stable form. Their boiling points and melting points are shown below:

Compound	NbF_5	$NbCl_5$	$NbBr_5$	TaF_5	$TaCl_5$	$TaBr_5$	TaI_5
M.p., °C	76	210	268	97	220	280	496
B.p., °C	229	254	362	230	239	349	543

All these compounds have the structure of a trigonal bipyramid, with the metal atom in the center. Niobium pentachloride is known in two forms: the white and the yellow form (transition point $-183°C$).

Tantalum pentachloride decomposes on being heated to give $TaCl_3$ and chlorine, after which the former reacts with excess $TaCl_5$ to give $TaCl_4$. Tantalum tetrachloride is a dark green solid, which is readily decomposed by water, according to the equation:

$$2TaCl_4 + 5H_2O \longrightarrow TaCl_3 + Ta(OH)_5 + 5HCl$$

Tantalum trichloride and niobium trichloride, which are black and dark green respectively, are obtained by thermal decomposition of the chlorides MCl_5. They dissolve in cold water, forming blue (niobium) or green (tantalum) solutions. Both solutions act as very strong reducing agents and are gradually oxidized by atmospheric oxygen. When alkalis are added to the solution of $TaCl_3$, the green $Ta(OH)_3$ precipitates out; it is so strongly reducing that, on being boiled with water, it decomposes it as follows:

$$Ta(OH)_3 + 2H_2O \longrightarrow Ta(OH)_5 + H_2$$

The dark green nonvolatile $TaCl_2$ may be obtained by thermal decomposition of $TaCl_3$ at 500°C. It is practically insoluble in water, but is oxidized in the cold, with evolution of hydrogen and formation of Ta^{3+} ions.

Halides of metals of the chromium sub-group /62/

Almost all the halides of metals of the chromium sub-group are high-melting crystalline substances. Products in which the oxygen in the trioxides MO_3 is fully substituted by halogen have only been obtained for molybdenum and tungsten. These metal fluorides are formed by direct interaction with fluorine. Molybdenum hexafluoride MoF_6, m.p. 18°C, b.p. 35°C and tungsten hexafluoride WF_6, m.p. 2°C, b.p. 18°C, are colorless, low-melting and very volatile. Organic solutions of WF_6 are intensely colored. Metal hexafluorides are very reactive. They are readily decomposed by water to the oxyfluorides MOF_4 and MO_2F_2. Other halides of MX_6 type are known for tungsten only. The dark violet WCl_6, m.p. 284°C, b.p. 337°C, is formed by direct reaction between elements in the heat. It has the structure of a regular octahedron. It is readily soluble in alcohol and ether, and practically insoluble in cold water, but is readily decomposed by hot water, with formation of $WOCl_4$ and WO_2Cl_2. The blue-black WBr_6 displays similar properties.

Few pentavalent compounds of metals in this sub-group have been studied. Tungsten pentachloride can be obtained by repeated distillation of WCl_6 in a stream of hydrogen, while $MoCl_5$ is obtained by heating metallic molybdenum powder in a stream of chlorine. Both WCl_5, m.p. 248°C, b.p. 276°C, and $MoCl_5$, m.p. 194°C, b.p. 268°C, form greenish-black crystals. The brown-violet WBr_6, m.p. 276°C, b.p. 333°C, is known. Neither solid nor fused molybdenum pentachloride conducts electric current. Its crystalline structure may be represented by two octahedra with a common edge. The bonds Mo—Mo are absent /95/. Both WCl_5 and $MoCl_5$ are decomposed by water with formation of the metal trichloroxide.

The tetravalent metal halides which have been prepared include only CrF_4, which is formed by the reaction between fluorine at 400 — 500°C with Cr or CrF_3. This compound is markedly volatile at temperatures as low as 150°C, deliquesces in moist air and etches glass. The reaction between $CrCl_3$ and chlorine at 700°C yields $CrCl_4$, which is stable in the gas phase only. Chlorides of tetravalent molybdenum and tungsten are the compounds which have been studied in most detail. The brown-colored $MoCl_4$, which is formed by the reaction between MoO_2 and chlorine in the heat, readily sublimes as yellow vapors, but the brown-gray WCl_4 is not volatile. It may be prepared as crystals by heating WCl_6 in a stream of hydrogen at a high temperature. Molybdenum tetrachloride displays a trigonal symmetry /95/. These compounds are very hygroscopic and are decomposed by water. Similar properties are displayed by the black-colored WI_4, while the reddish-brown WF_4 is more stable with respect to water.

The trivalent halides of the metals in this sub-group are cheapest and are most easily prepared. Anhydrous chromium chloride $CrCl_3$ is formed by direct reaction between the elements in the heat. It forms red-violet crystals, which are readily distilled in a stream of chlorine. Chromium trichloride $CrCl_3$ has an m.p. of 1150°C and is practically insoluble in water. However, in the presence of traces of $CrCl_2$ or any other reducing agent the dissolution is rapid and proceeds with a marked evolution of heat. Two varieties of chromium trichloride hexahydrate $CrCl_3 \cdot 6H_2O$ may be formed, namely, the dark green and the violet modification, of which the dark green form is much more strongly hydrolyzed than the violet form. The green CrF_3 (subl. pt. 1200°C) and the almost black $CrBr_3$ are very close in their properties to $CrCl_3$; the triiodide cannot be isolated free, but the violet-colored hydrate $CrI_3 \cdot 9H_2O$ is known.

Unlike for chromium, the trivalent state is not typical of molybdenum and tungsten. Molybdenum trichloride may be prepared by heating $MoCl_5$ at 250°C in a stream of hydrogen. It is a dark red crystalline substance, insoluble in water and in hydrochloric acid.* The black-colored bromide of a similar composition may be prepared directly from the elements, while the fluoride is prepared by reacting $MoBr_3$ with HF at 600°C. The chloride of W(III) is not known in the free state, but only as greenish-yellow double salts of the type $2WCl_3 \cdot 3MCl$, e.g., $2WCl_3 \cdot 3KCl$.

The chloride of bivalent chromium $CrCl_2$ is formed by the reaction between the metal and hydrochloric acid in an atmosphere of hydrogen. It may be prepared by heating metallic chromium in a stream of gaseous hydrogen chloride or by reducing $CrCl_3$ by hydrogen at about 600°C. Anhydrous $CrCl_2$ is a colorless crystalline substance, m.p. 824°C, very hygroscopic, which dissolves in water to give deeply colored solutions. It may be isolated from such solutions as the blue-colored $CrCl_2 \cdot 4H_2O$ hydrate, which isomerizes to a green form above 38°C and at 51°C is converted to the blue-colored trihydrate.

All halides of bivalent molybdenum and tungsten are known except for the fluorides. The yellow $MoCl_2$ may be prepared by heating molybdenum in phosgene vapors. It is almost insoluble in water, but is soluble in alcohol and ether, in which it is trimeric. The orange-colored $MoBr_2$ displays properties similar to those of the chloride. The same applies to the brown-colored molybdenum iodide. The gray WCl_2, which is unstable in the air, may be prepared by heating WCl_4 in a stream of dry gaseous CO_2. It is a strong reducing agent and reacts with water with a vigorous evolution of hydrogen. The bromide WBr_2 displays similar properties, while the brown WI_2 is practically insoluble in cold water, while being decomposed by hot water.

Metal alkoxides and aryloxides

Titanium alkoxides are the most important alkoxides of transition metals employed as components of complex polymerization catalysts. We shall accordingly limit our discussion to the derivatives of titanium.

* It has been recently established /95/ that $MoCl_3$ crystals may exist in two (α- and β-) modifications, which differ in the mode of packing of the anions.

Titanium alkoxides and aryloxides

These compounds may belong to one of two types: titanium alkoxyhalides (aryloxyhalides) and titanium tetraalkoxides (tetraaryloxides).

Up till now, titanium alkoxychlorides and titanium tetraalkoxides have been studied in most detail. Other titanium derivatives have not yet been studied in detail.

Titanium alkoxychlorides are low-melting, colorless crystalline substances or viscous liquids. The viscosity of titanium trialkoxychlorides decreases with increasing molecular weight. Titanium alkoxybromides are yellowish-colored substances. Titanium aryloxides are red-colored solid substances, with much higher melting points. All these compounds are very hygroscopic and are as a rule readily soluble in organic solvents. Titanium alkoxychlorides with straight chain alkyl groups are stable on storage and may be vacuum-distilled, which results in only a small degree of disproportionation. Titanium alkoxychlorides with secondary and tertiary alkyl carbons are unstable and decompose on being stored or heated to give titanium oxychloride $TiOCl_2$, alkyl chloride RCl, hydrogen chloride and polymerization products /65/.

Titanium tetraalkoxides are, save for a few exceptions, colorless or yellowish liquids, with high boiling points. Titanium tetraaryloxides are red-colored, solid substances. All these compounds, just like titanium alkoxy halides, are readily soluble in most organic solvents. The molar refraction of titanium tetraalkoxides is 20.63 ± 0.06 /66/, while the refraction of the Ti — O-bond in alkoxy derivatives of titanium is 4.08 /67/. Spectroscopic studies of esters of orthotitanic acids are available /68, 69/.

Lower titanium tetraalkoxides, up to $(C_5H_7O_4)Ti$, are stable to heat and may be distilled under atmospheric pressure without decomposition. As the length of the alkyl radical increases, the compound becomes less stable to heat. However, even lower titanium tetraalkoxides, while being stable to heat, undergo condensation on prolonged heating, which is accompanied by the evolution of volatile substances and formation of polymeric compounds:

$$-\underset{OR}{\overset{OR}{\underset{|}{\overset{|}{Ti}}}}-O-\underset{OR}{\overset{OR}{\underset{|}{\overset{|}{Ti}}}}-O-$$

The alkoxy group is readily split off when $Ti(OR)_4$ is acted upon by water:

$$-\underset{|}{\overset{|}{Ti}}-OR + H_2O \longrightarrow -\underset{|}{\overset{|}{Ti}}-OH + ROH$$

If the amount of water introduced into the reaction mixture is insufficient to ensure a full hydrolysis of the ortho-ether, an intermolecular condensation of hydrolysis products takes place, with formation of polytitanoxane compounds:

$$2-\underset{|}{\overset{|}{Ti}}-OH \longrightarrow -\underset{|}{\overset{|}{Ti}}-O-\underset{|}{\overset{|}{Ti}}- + H_2O$$

Reaction with alcohol results in trans-etherification:

$$Ti(OR)_4 + 4R'OH \rightleftarrows Ti(OR')_4 + 4ROH$$

Titanium tetraalkoxides react with hydroxycarbonyl and enolized compounds (acetylacetone, acetoacetic ester, etc.), when orthotitanic acid esters with a chelate structure are formed. Thus, the reaction between titanium tetraalkoxides and acetylacetone yields two different products, depending on the initial reactant ratio /4/:

```
                 CH3        H3C                        CH3
             O—C               C=O.   OR   .O—C
(RO)3—Ti          CH  and  HC          Ti          CH
             O=C               C—O    OR    O=C
                 CH3        H3C                        CH3
```

These compounds are stable to water. Similar compounds may be prepared by the reaction between titanium halides and acetylacetone /70, 71/. Some of the physical properties of titanium tetraalkoxides and tetraaryloxides are listed in Table 8.

TABLE 8. Certain physical properties of the most important titanium tetraalkoxides and tetraaryloxides /72/

Formula	B.p., °C (mm Hg)	M.p., °C	Formula	B.p., °C (mm Hg)	M.p., °C
$C_2H_5OTiCl_3$	185 — 186	81 — 82	$(C_2H_5O)_2TiBr_2$	95 — 105 (5)	47 — 50
iso-$C_3H_7OTiCl_3$	65 (1)	78 — 79			(decomposes)
$(C_2H_5O)_2TiCl_2$	142 (18)	40 — 50	$(C_2H_5O)_4Ti$	236 — 237	40
		(decomposes)	(iso-$C_3H_7O_4$)Ti	920 (760)	—
(iso-$C_3H_7O)_2TiCl_2$	160 (18)	—		97 (7.0)	
$(C_2H_5O)_3TiCl$	176 (18)	—	$(C_6H_5O)_4Ti$	267 (3)	153 — 154

The methods of synthesis of orthotitanic acid esters are based on the substitution of the chlorine atoms in $TiCl_4$ by alkoxy or aryloxy groups, and on the substitution of the alkoxy groups in titanium tetraalkoxides by other alkoxy groups or by aryloxy groups.

Shiihara et al. /79/ gave the most detailed review of the various derivatives of titanium. Here we shall merely say a few words about the most frequently used methods for the preparation of titanium tetraalkoxides. Titanium tetrachloride reacts with alcohols up to the stage of formation of dialkoxydichlorotitanium compounds /80, 81/. In order to substitute all four chlorine atoms by alkoxy groups, the hydrogen chloride evolved during the reaction must be bound by ammonia /73, 74/, an organic base /73, 76/ or sodium alcoholate in lieu of alcohol /73/. The best method for the preparation of titanium ortho-ethers is the reaction between alcohol and the ammoniate of titanium tetrachloride $TiCl_4 \cdot 8NH_3$, which is prepared by passing ammonia into a solution of $TiCl_4$ in cyclohexane /77/:

$$TiCl_4 \cdot 8NH_3 + 4ROH \longrightarrow Ti(OR)_4 + 4NH_4Cl + 4NH_3$$

In order to avoid the need for filtering the reaction mixture, it has been suggested /78/ that the reaction be conducted in formamide or in adiponitrile, in which NH_4Cl is soluble, while $(RO)_4Ti$ precipitates out.

2. METAL ACETYLACETONATES

Most metal acetylacetonates are crystalline powders which are readily soluble in organic solvents.

It was found by Cox /82/ that acetylacetonates of bivalent metals are spiranlike compounds with a planar arrangement of the rings:

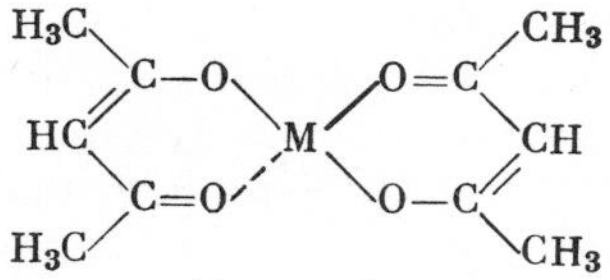

```
H3C                      CH3
   \C—O\        /O=C/
HC//     \M /         \CH
   \C=O/   \O—C//
H3C/                     \CH3
```

M — metal

Acetylacetonates of higher valency metals have a similar structure. It has been shown /83, 84/ that the stability of the M — O-bond in these compounds depends on the nature of the metal and increases with increasingly electronegative character of the metal in the following sequence:

$$Ba \longrightarrow Sr \longrightarrow Ca \longrightarrow Mg \longrightarrow Cd \longrightarrow Mn \longrightarrow Pb \longrightarrow Zn \longrightarrow Co \longrightarrow$$
$$\longrightarrow Ni \longrightarrow Fe \longrightarrow Cu \longrightarrow Be \longrightarrow Hg$$

All methods of preparation of metal acetylacetonates can be grouped under the following five headings:

1. Reaction between metal hydroxide and acetylacetone /85, 86/:

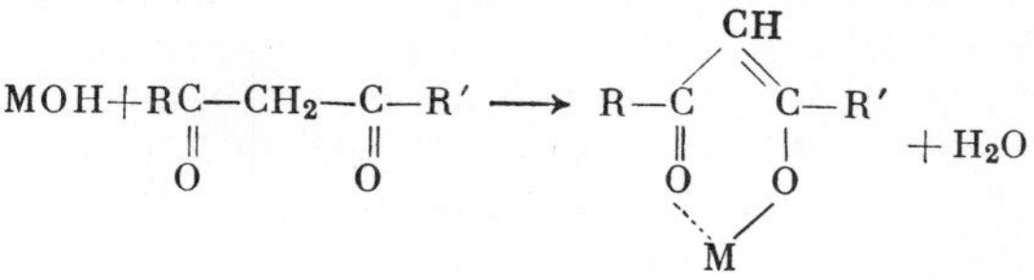

$$MOH + RC(=O)—CH_2—C(=O)—R' \longrightarrow R—C(=O\cdots M—O—)=CH—C—R' + H_2O$$

The reaction proceeds at room temperature. In this way cobalt, nickel, iron and copper acetylacetonates are readily prepared.

2. Reaction between sodium acetylacetonate and a metal salt /87/:

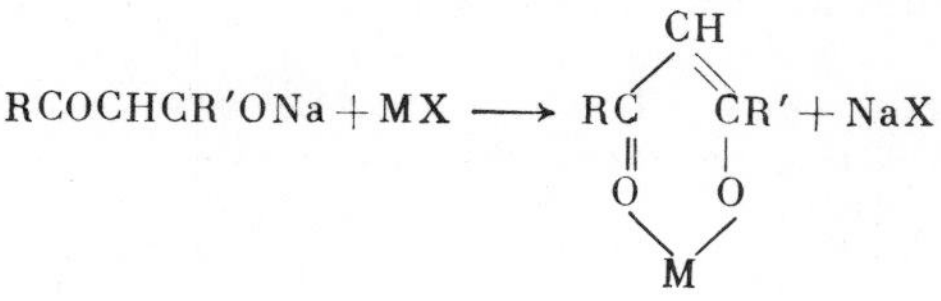

$$RCOCHCR'ONa + MX \longrightarrow RC(=O—M—O—)=CH—CR' + NaX$$

Acetylacetonates of Al, Fe, Ni, Cu, Mn, Cr, etc. have been obtained by this method.

3. Reaction between acetylacetone with metal salts in the presence of ammonium hydroxide /88/:

$$MX_2 + 2NH_4OH + 2RCOCH_2COR \longrightarrow (RCOCHCOR)_2M + 2NH_4X + 2H_2O$$

The reaction probably takes place via an intermediate compound; the ammonium salts of the diketone is formed first, and reacts with the metal salt to give the chelate compound.

4. Reaction between acetylacetone and nascent metal ions /89/.

In this way it is possible to prepare acetylacetonate of trivalent manganese by reacting stoichiometric proportions of acetylacetone, manganous sulfate and potassium permanganate:

$$6MnSO_4 + 2KMnO_4 + 18CH_3COCH_2COCH_3 \longrightarrow 6Mn(C_5H_7O_2)_3 + 2MnO_2 + K_2SO_4 + 5H_2SO_4 + 4H_2O$$

5. Reaction of acetylacetone with metal oxides /90, 91/:

$$2RCOCH_2COR + MeO \longrightarrow (RCOCHCOR)_2M + H_2O$$

Derivatives of copper and lead have been obtained in this way.

The preparation and certain properties of a number of individual metal acetylacetonates in most frequent use are described below.

Aluminum acetylacetonate. This compound is prepared from aluminum chloride, ammonia and acetylacetone, dissolved in absolute alcohol. The reaction mixture is stirred for 2 hours, NH_4Cl is filtered off and ethanol is expelled by distillation from the ethanolic solution of aluminum acetylacetonate /92, 93/.

It is also possible to prepare the compound in a similar manner from a basic salt of aluminum and acetylacetone or from

$$Al_2(SO_4)_3,\ NH_3 \text{ and } CH_3{-}\overset{O}{\overset{\|}{C}}{-}CH_2{-}\overset{O}{\overset{\|}{C}}{-}CH_3 \text{ /94/.}$$

Titanium acetylacetonate. The reaction between equimolecular amounts of acetylacetone and titanium tetraalkoxide yields a pentacoordinated titanium compound /68/:

$$(RO)_3Ti(O{-}C(CH_3){=}CH{-}C(CH_3){=}O) \quad (R{=}C_2H_5,\ C_3H_7,\ C_4H_9)$$

If two equivalents of the diketone are employed, the titanium atom becomes hexacoordinated. Solutions of these compounds in benzene are monomeric.

Titanium trichloride reacts with acetylacetone to form a compound with M. W. 490 — 540, to which the following structure is ascribed /96/:

$$\left[\begin{array}{l} H_3C\diagdown \\ \quad C{-}O{-} \\ HC\!\!\diagup\!\!\!\!/ \\ \quad C{=}O \\ H_3C\diagup \end{array}\right] \diagdown Ti \begin{array}{c} Cl \\ \\ Cl \end{array} Ti \diagup \left[\begin{array}{r} \diagup CH_3 \\ O{=}C \\ CH \\ {-}O{-}C \\ \diagdown CH_3 \end{array}\right]_2$$

The following interesting syntheses are known /97/:

$$3TiCl_4 + 6CH_3COCH{=}C(OH)CH_3 \longrightarrow 6HCl + 2[Ti(C_5H_7O_2)_3]^+ + (TiCl_6)^{2-}$$

$$TiCl_4 + 3CH_3COCH{=}C(OH)CH_3 + FeCl_3 \longrightarrow 3HCl + [Ti\,(C_5H_7O_2)_3]^{2+} + (FeCl_4)^-$$

It is considered that the products contain hexacoordinated titanium. bis-(Acetylacetonate)titanium dichloride is formed by the reaction between titanium tetrachloride and acetylacetone; the compound is monomeric in boiling benzene /98/. As a result of the reaction with isopropanol, the chlorine atoms are substituted by isopropoxy groups. The compound $Ti(C_5H_7O_2)_3$ can be prepared by the reaction between titanium trichloride and acetylacetone in the presence of dry ammonia in benzene solution /99/. The NMR spectrum of Ti(IV) acetylacetonate has also been taken /180/.

Zirconium acetylacetonate. Well-shaped crystals of the decahydrate of $Zr(C_5H_7O_2)_4$ were obtained by reacting acetylacetone with an aqueous solution of zirconium nitrate or zirconium chloride /100, 101/. The following method of preparation of zirconium tetraacetylacetonate is convenient.

Zirconium tetrachloride octahydrate (5.8 g) is dissolved in 50 ml water and the solution is cooled to 15°C. Fifty ml of a 10% solution of sodium carbonate and 10 g of acetylacetone are slowly introduced into another vessel, with stirring. This solution is cooled to 0°C, filtered if necessary, and poured into the solution of zirconium chloride, which should also be cooled to 0°C. The precipitation of zirconium acetylacetonate takes about one hour. The precipitate is filtered off, washed with cold water and dried. It contains insignificant amounts of zirconium hydroxide as impurity. To obtain the pure compound, the crude product is dissolved in benzene (5 g substance in 25 ml benzene), the solution is filtered and petroleum ether is added to the filtrate, when pure anhydrous zirconium acetylacetonate precipitates out.

Hafnium acetylacetonate may be prepared in the same manner.

Zirconium acetylacetonate is a colorless crystalline substance, which forms typical monoclinic, doubly refracting needles /101/; the density at 25°C is 1.415 g/cm^3 /100/. The compound melts at 194 — 195°C, boils at 82°C / 0.001 mm Hg and begins to decompose at about 125°C; its Raman spectrum has an absorption band at 1520 cm^{-1} /102/; its water solubility is about 10 g in 100 ml solution at 20°C; it is insoluble in acetone and does not react with absolute alcohol below 40°C.

Vanadium acetylacetcnate is prepared from VCl_4 and acetylacetone or sodium acetylacetonate /103/. Vanadium (III) acetylacetonate and vanadyl acetylacetonate may be prepared in a similar manner. Vanadyl acetylacetonate forms blue-green, monoclinic crystals, m.p. 250°C. Vanadium(III) acetylacetonate is a brown-colored crystalline substance, m.p. 182°C, which is easily oxidized to the vanadyl derivative.

Chromium acetylacetonate is prepared by interaction between chromium chloride and acetylacetone in the presence of sodium carbonate /104/. It forms red-violet monoclinic crystals, m.p. 216°C, which are soluble in most organic solvents.

Molybdenum acetylacetonate is formed by reacting molybdenum compounds with acetylacetone /105, 106/.

Manganese acetylacetonate is obtained by the reaction between manganese salts and acetylacetone and sodium carbonate, or by treatment of the salt with aluminum acetylacetonate in an organic solvent /107/. It is a brown-colored, monoclinic, crystalline product, m.p. 172°C, soluble in organic solvents, insoluble in water and stable in the air.

Ferric acetylacetonate is formed by reacting acetylacetone with an aqueous solution of $Fe_2(SO_4)_3$ and sodium carbonate, or else by reacting ferric chloride with sodium acetylacetonate /108/. It may also be prepared by reacting aluminum acetylacetonate with a solution of an iron salt in absolute alcohol /107/. The corresponding derivatives of ferrous iron are prepared in a similar manner, using ferrous sulfate or ferrous chloride. Ferric acetylacetonate forms yellowish-red prisms, m.p. 182°C, which are readily soluble in organic solvents, but sparingly soluble in water. Its aqueous solution is practically nonconducting /109/. It is decomposed by alkalis, the colloidal $Fe(OH)_3$ being formed on boiling. It is stable to acids.

Cobalt(II) acetylacetonate may be prepared from $Co(OH)_2$ and acetyl-acetone. Cobalt acetylacetonate forms red crystals, which sublime. It is soluble in water and in most organic solvents /110/. Cobalt(III) acetyl-acetonate is formed by oxidizing the corresponding cobalt(II) derivative with sodium hypochlorite or hydrogen peroxide. The oxidized product separates out as dark green, monoclinic crystals, m.p. 241°C, which are readily soluble in organic solvents.

Nickel acetylacetonate is prepared from ammonium acetylacetonate and nickel oxide or a nickel salt /107/. The compound may also be prepared from $Ni(OH)_2$ or $NiCO_3$ and acetylacetone by grinding the salts with acetylacetone in a special apparatus /109/. Other workers described the preparation of nickel acetylacetonate from acetylacetone and a basic solution of a nickel complex /111/, or a nickel salt and ammonia /105/. Nickel acetylacetonate forms pale green crystals, m.p. 228°C, soluble in water and in some organic solvents, such as alcohol and heptane. It boils at 220°C/11 mm Hg /112/.

3. ORGANIC DERIVATIVES OF METALS

Despite repeated attempts, it has not yet been possible to prepare organic derivatives of most of the transition metals. The reason for this is to be sought in the weakness of the covalent bond between the metal and the organic radical /113, 114/. Nevertheless, these attempts have not been abandoned. In the early 1950's considerable progress could be made in the development of organic chemistry of transition metals. This was due, on one hand, to the discovery (in 1951) of dicyclopentadienyl derivatives of transition metals, which are organometallic compounds of unusual structure and properties.

Its first representative — dicyclopentadienyliron, later named ferrocene — was first prepared in 1951 /115/. Soon afterwards, dicyclopentadienyl derivatives of almost all transition metals were prepared, as well as diindenyl compounds, cyclopentadienylcarbonyls and cyclopentadienyl-nitrosyls of a number of metals, the properties and structure of which have been described in detail /116 — 120/.

On the other hand, owing to the discovery of complex organometallic catalysts, interest has arisen in the chemistry of alkyl and aryl derivatives of transition metals.

Finally, despite the fact that benzene derivatives of chromium were discovered as early as 1919 /121/, the arene derivatives of transition metals were only recognized following the experimental and theoretical work of Zeiss et al. /122, 123/.

In what follows we shall discuss only the most common modes of preparation and properties of cyclopentadiene compounds of transition metals, while their alkyl and aryl derivatives will be treated in more detail.

Cyclopentadienyl derivatives of transition metals

Cyclopentadienyl compounds of transition metals are mainly prepared in the following two manners:

1) by the reaction between the metal salt (usually acetylacetonate) and magnesium, lithium or sodium cyclopentadienyl:

$$TiX_4 + 2NaC_5H_5 \longrightarrow Ti(C_5H_5)_2X_2 + 2NaX$$

2) by direct reaction between cyclopentadiene and a metal salt or metal carbonyl.

The former method was employed to prepare the derivatives of cobalt /124/, titanium, zirconium and vanadium /125, 126/, nickel /127/, manganese /128/, molybdenum /129/ and iron /130/.

The reaction between cyclopentadienylmagnesium bromide with metal acetylacetonates is successfully employed if the halide is not readily soluble in ether /125/. Cyclopentadienyl compounds of iron /131/, nickel /125/, cobalt /132/, ruthenium /133/, rhodium and iridium /134/ have been prepared in this way.

Reaction between cyclopentadienylsodium or lithium and alkyl halides gave high yields of bis-cyclopentadienyl compounds of iron /135/, titanium and vanadium /126/, molybdenum /129/, tantalum /126/ and manganese /128/. The method is also suitable for the preparation of tricyclopentadienyl compounds of scandium, yttrium and lanthanides /136, 137/. Tetrahydrofuran or ethylene glycol dimethyl ether are used as solvents. Salts of trivalent bis-cyclopentadienyltitanium $(C_5H_5)_2TiX$ were obtained by electrometric reduction of $(C_5H_5)_2TiX_2$ /126/ or by interaction with lithium aluminum hydride /138/.

Physical and chemical properties. All cyclopentadienyl compounds of Group VIII metals have an antiprismatic structure, with a center of symmetry. The cyclopentadienyl ring in these compounds is bound to the metal as in ferrocene, for which a pentagonal antiprism "sandwich" structure was proposed in 1952 /139/. The iron is located in the symmetry center of the molecule, which has only one type of C — H bond:

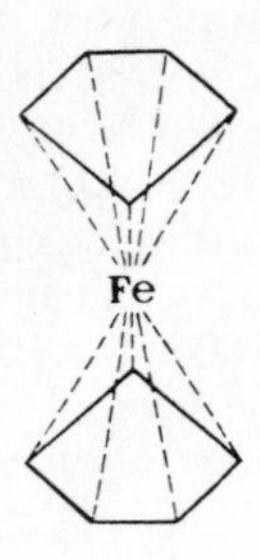

The "sandwich" structure of ferrocene was subsequently confirmed by quantum-mechanical calculations, by the method of molecular orbitals /140 — 144/, by the study of magnetic properties /139, 145/, X-ray and spectroscopic methods /140, 146, 147/. Dyatkina /148/ published an extensive review on the structure and nature of the bonds in aromatic metal complexes.

The salts of tetravalent bis-cyclopentadienyltitanium $(C_5H_5)_2TiX_2$ are diamagnetic /126, 136/, stable in the air and soluble in organic solvents. They are soluble in water, with some small decomposition /149/. bis-Cyclopentadienyltitanium chloride reacts with phenyllithium to form diphenylcyclopentadienyltitanium, which gradually decomposes at room temperature /150, 151/:

$$(C_5H_5)_2TiCl_2 + 2C_6H_5Li \longrightarrow (C_5H_5)_2Ti(C_6H_5)_2 + 2LiCl$$

The bis-cyclopentadienyl compound of bivalent titanium $(C_5H_5)_2Ti$ is diamagnetic, and thermally unstable; it decomposes on being heated in an inert gas atmosphere below the melting point (about 130°C). It is also decomposed in the air /152/. It reacts very slowly with water. It reacts with dilute hydrochloric and sulfuric acids to give a compound of tetravalent titanium, and forms an etherate with tetrahydrofuran. Solutions of bis-cyclopentadienyltitanium in ammonia are practically nonconducting.

bis-Cyclopentadienyldibromozirconium is diamagnetic. Its other properties have not been studied.

bis-Cyclopentadienyldichlorovanadium is paramagnetic; it dissolves in polar organic solvents and forms green solutions in water. It is stable in acid media for several hours. It is reduced by lithium aluminum hydride to $(C_5H_5)_2V$ /153/. The bis-cyclopentadienylvanadium ion of trivalent vanadium $(C_5H_5)_2V^+$ is oxidized by atmospheric oxygen /126/. bis-Cyclopentadienylvanadium is paramagnetic and is stable to water and even to acids. X-ray studies indicate that these compounds have a "sandwich" structure /154/. The IR spectra of bis-cyclopentadienyl compounds of V, Nb and Ta resemble one another and display absorption bands typical of the "sandwich" structure /126/. It may be assumed that the chemical properties of these compounds are essentially similar.

bis-Cyclopentadienylchromium $(C_5H_5)_2Cr$ is paramagnetic /155/, stable to heat up to 300°C, but is very readily oxidized in the air. When finely comminuted, it spontaneously bursts into flame.

bis-Cyclopentadienylmanganese is paramagnetic /155/. Its structure resembles that of ferrocene, but the bond between the manganese and the cyclopentadienyl rings is ionic /156, 157/. On being heated, the brown-colored bis-cyclopentadienylmanganese changes color and passes into another modification /135, 137/. It decomposes as a result of oxidation.

bis-Cyclopentadienylcobalt $(C_5H_5)_2Co$ is very readily oxidized to the very stable diamagnetic cation $(C_5H_5)_2Co^+$, which gives water-insoluble salts; these salts are not decomposed by sulfuric and nitric acids. In anhydrous solvents $(C_5H_5)CoBr$ is reduced by lithium aluminum hydride to $(C_5H_5)_2Co$ in a high yield /153/.

The bond between the metal and the cyclopentadiene ring in $Co(C_5H_5)_2$ is not ruptured by the action of water, but the compound is merely slowly oxidized to the cation /135/. The reaction between $(C_5H_5)_2Co^+$ and cyclopentadienylsodium yielded the compound $Co_2(C_5H_5)_5$, which was stable to hydrolysis and to which the structure $(C_5H_5)_2Co(C_5H_5)Co(C_5H_5)_2$ was ascribed. Its dipole moment is zero /158/. The density of $(C_5H_5)_2Co$ at 18°C is 1.49 g/cm^3.

bis-Cyclopentadienylnickel $(C_5H_5)_2Ni$ is paramagnetic, green-colored, can be vacuum-distilled and is readily soluble in nonpolar solvents. It is gradually oxidized in the air, especially when in solution. It is insoluble in water and does not react with it. Its density is 1.47 g/cm^3 at 18°C /159/.

Alkyl and aryl derivatives of transition metals

Stable covalent organometallic compounds, except for alkylated derivatives of tetravalent platinum and trivalent gold, have only been prepared comparatively recently. Most halides of transition metals react with metal alkyls and aryls, but very often the products cannot be isolated pure.

Best results were obtained by careful choice of the experimental conditions or by formation of definite structures corresponding to electronic stability. This is achieved if the transition metal assumes the electronic configuration of the following inert gas or if it loses 2 — 4 electrons /160/.

Compounds of transition metals with σ-bound alkyl and aryl groups are here considered in accordance with their position in the periodic table. Reviews on the subject are available /79, 118, 160/; we shall accordingly limit ourselves to individual compounds of transition metals which are most often employed in the preparation of stereospecific catalysts.

Alkyl (aryl) derivatives of Group IV transition metals*

Titanium derivatives of this type have been studied in most detail. The starting point in the development of the chemistry of organotitanium compounds, which had hitherto been thought to be nonexistent, were the numerous studies of various catalytic systems consisting of titanium halides and metal alkyls. Subsequent to the preparation and identification of the first alkyltitanium halide /161/, a large number of compounds of the type R_nTiX_{4-n} were prepared, where $n = 1, 2, 3, 4$, R is methyl, ethyl or butyl and X is Cl, Br or I.*

Monoalkyl and dialkyl compounds of titanium have been studied in detail. They are prepared by exchange reaction between titanium halide and

* The first attempts to isolate and identify a compound with Ti — C bond were made by Pittser /162/.

organoaluminum derivatives of the general formula R_nAlX_{3-n}, in an inert atmosphere at temperatures varying between − 80 and 0°C. Hexane, benzene and other hydrocarbons are used as solvents. In order to obtain the mono-alkyl derivative, the reaction is conducted in the presence of excess TiX_4. Dialkyltitanium halides are obtained if $TiX_4:R_2AlX = 1:2$. The alkyltitanium halide formed is isolated by vacuum distillation of the reaction mixture, after converting the alkylaluminum compound into an insoluble or nonvolatile state with the aid of alkali metal halides or other complex-forming substances /163, 164/. Alkyltitanium halides and even tetraalkyltitanium compounds may also be prepared if various other metal alkyls such as methyllithium, phenyllithium /165, 166/, dimethylzinc /167/, diphenylmercury /168/, tetraethyllead /169/, etc., are used as alkylating agents.

All organotitanium compounds are unstable. In the series R_nTiX_{4-n}, where R is alkyl or aryl, compounds with a smaller number of organic groups and with a more negative X group are the more stable /166/. The stability of methyl derivatives of titanium to heat has been studied in detail; it was shown that, in the absence of moisture and oxygen, methyltitanium trichloride is stable at 20°C for several days.

When the temperature is increased to 100°C, CH_3TiCl_3 decomposes with formation of $TiCl_3$, methane and polymethylene. The $TiCl_3$ thus formed acts as the catalyst of the decomposition /163/. Tetramethyltitanium is stable at −70°C; when the temperature is raised to room temperature value, spontaneous decomposition occurs with formation of a black precipitate. The decomposition products are pyrophoric /170/. A number of phenyl derivatives behave in a similar manner. Thermal decomposition of tetraphenyltitanium yields diphenyltitanium, which is much more stable to heat than tetraphenyl-titanium. It decomposes in vacuo at 200 — 250°C, with the evolution of metallic titanium /168/. The thermal decomposition of tetramethyltitanium has been studied in most detail /172/.

The most thoroughly studied reactions of organotitanium compounds with Ti — C σ-bonds are oxidation, hydrolysis and alcoholysis. The oxidation of lower valency titanium derivatives is vigorous, and is often accompanied by spontaneous combustion /168, 170, 171/. In the series of derivatives of tetravalent titanium the stability to oxygen increases when cyclopentadienyl groups are introduced into the molecule. Alkoxy derivatives of titanium are formed in this manner:

$$CH_3TiCl_3 + {}^1/_2O_2 \longrightarrow CH_3OTiCl_3$$

Even though bis-cyclopentadienyltitanium compounds $(C_5H_5)_2TiR_2$ are more stable to oxygen, they are oxidized in solution at a fast rate, forming the corresponding alkoxy derivatives. The oxidation of diphenyl-bis-cyclo-pentadienyltitanium in benzene at 50 — 60°C showed that the "sandwich" group $(C_5H_5)_2Ti$ is not oxidized. This compound is, however, fully oxidized by aqueous hydrogen peroxide at 50°C, when CO and CO_2 are formed /172/.

In compounds of the type CH_3TiCl_3, $(CH_3)_2TiCl_2$ and other alkyl halides the σ-bond Ti — C is broken by water and alcohol, with the formation of alkane. However, molecules containing cyclopentadienyl radicals such as $(C_5H_5)_2Ti(CH_3)X$ are stable to water and are decomposed only by dilute acids /173/. Phenyl derivatives of titanium react with water similarly to alkyl compounds of titanium /174/.

Alkyl and aryl derivatives of zirconium and hafnium are as yet almost unknown. However, it was pointed out by Razuvaev and Latyaeva /118/ that formation of organic compounds of these metals by exchanging halogen against an [alkyl] radical is possible. The preparation of $Zr(C_5H_{11})_2$ from metal hydrides and diazo compounds at low temperatures was recently reported /175/. Zirconium tetrachloride was also reported to react with diethylmercury and diphenylmercury /176/ and with organolithium and organomagnesium compounds /177/.

The following alkyl and aryl derivatives of titanium should be mentioned.

Methyltitanium trichloride — a dark violet, crystalline compound, m.p. 28.5°C, b.p. 37°C / 1 mm Hg. The compound melts to form a yellow liquid, which is soluble in many organic solvents. It is monomolecular in benzene solution /163/.

Tetramethyltitanium — lustrous yellow crystals, which can be distilled with ether in vacuo not above 0°C, stable at −78°C. Separates out of hexane solutions as crystals at −30 to −50°C /170/.

Trimethyltitanium iodide — yellow crystalline needles. Stable at low temperatures in the absence of oxygen /178/.

Phenyltriisopropoxytitanium — a practically colorless compound, m.p. 88 — 90°C, stable in inert atmosphere in the absence of moisture /179/. The compound contains a phenyl-titanium bond, as shown by the reaction with sublimate which yields phenylmercuric chloride, and also by oxidation to the phenoxy derivative.

Tetraphenyltitanium — a compound which is unstable to heat. It decomposes above −30°C with formation of diphenyl and diphenyltitanium.

Diphenyltitanium — a black, pyrophoric substance, which is soluble in tetrahydrofuran and benzene. More stable to heat than its tetravalent analog.

Methyl-bis-cyclopentadienyltitanium chloride — red-orange crystals, m.p. 168 — 169°C, soluble in water in the cold.

Dimethyl-bis-cyclopentadienyltitanium — orange-colored crystalline needles with a typical odor. Stable to water and oxygen, but rapidly decomposes in the light, with darkening.

Trimethylcyclopentadienyltitanium — yellow compound, stable at −70°C, distils at −20°C / 0.5 mm Hg /171/. At −80°C the compound may be isolated from solution in pentane as lemon-yellow crystals, which are pyrophoric in the air at room temperature /173/.

Alkyl (aryl) derivatives of transition metals in other groups

Up till recently, the phenyl derivatives were the only known organic compounds of metals in Group V with an M — C σ-bond. The early studies quoted by Cotton /181/ contained no convincing evidence in favor of the existence of alkyl or aryl derivatives of these metals. The synthesis and isolation of a complex which contained diphenylvanadium was first reported in 1960 /182/. This complex $4C_6H_5Li \cdot V(C_6H_5)_2 \cdot 3.5(C_2H_5)_2O$ is very sensitive to atmospheric oxygen and to moisture. It loses ether at 50°C in vacuo, and decomposes above 80°C with the liberation of diphenyl. It was also shown /118/ that diphenylmercury in a solution of cyclohexane reacts under conventional conditions with vanadium and vanadyl halides as follows:

$$(C_6H_5)_2Hg + VOCl_3 \rightleftarrows C_6H_5VOCl_2 + C_6H_5HgCl$$

$$(C_6H_5)_2Hg + VCl_4 \rightleftarrows C_6H_5VCl_3 + C_6H_5HgCl$$

The formation of phenyl derivatives of vanadium has been confirmed by spectroscopic data. However, the compounds could not be isolated pure owing to their instability to heat. The exchange of phenyl radicals was also noted in the reaction between tetraphenyltin and VCl_4, but under more drastic conditions /183/. Phenyl derivatives of vanadium can be prepared by exchanging the cyclopentadienyl group in $C_5H_5V(CO)_4$ against the phenyl radical by reacting benzene, cyclopentadienyltetracarbonylvanadium, oxygen and HCl /184/. When the halogen in π-$(C_5H_5)_2VCl$ was exchanged against a phenyl group by reaction with 1 mole of phenyllithium, the compound $(C_5H_5)_2VC_6H_5$ could be isolated pure /185/ as black crystals, m.p. 92°C. Solutions of phenyl-bis-cyclopentadienylvanadium are stable for one week at room temperature in the absence of oxygen.

The first alkylvanadium compounds were obtained as mixed cyclopentadienyl compounds of the type $(C_5H_5)_2VR$ by the addition of alkyl halides to dicyclopentadienylvanadium, and also by the action of LiR on dicyclopentadienyldichlorovanadium. The methyl compound, m.p. 80 — 100°C, and the much less stable ethyl compound have been prepared in this manner /231/.

Little is known about organic derivatives of niobium and tantalum /181, 186/. A complex has been obtained /187/ as a result of the reaction between tetraphenylniobium and phenyllithium.

The complex $Nb(C_6H_5)_4 \cdot 3C_6H_5Li \cdot 3(C_2H_5)_2O$ is a black-violet crystalline substance, which is pyrophoric when dry. The ether is removed at room temperature, under a high vacuum. The compound is not volatile, is stable up to 40 — 50°C, and is readily oxidized and hydrolyzed.

Trialkyl derivatives of niobium and tantalum — $Nb(CH_3)_3Cl_2$ and $Ta(CH_3)_3Cl_2$ — have recently been obtained by alkylation of the metal pentachlorides with diethylzinc in pentane at −78°C. They are yellow crystalline compounds, which are stable at −78°C in the absence of water and oxygen; at room temperature they decompose within a few hours /201/.

Alkyl derivatives of Group VI metals have not yet been prepared pure. They were mentioned on one occasion in connection with the reaction between methyllithium and chromous chloride, when it would seem that dimethylchromium was obtained. Chromic chloride probably yielded trimethylchromium /188/. Cyclopentadienyl derivatives of chromium with methyl and ethyl groups in the molecule have been isolated. They are crystalline or oily substances, which are diamagnetic, can be distilled in high vacuum and are readily soluble in organic solvents. They are decomposed slowly in the air /189, 190/. Aryl compounds of chromium seem to have been known as early as the 1920's. However, a detailed study of the compounds obtained by the reaction between phenylmagnesium bromide and redistilled anhydrous chromic chloride at −10°C showed that these are not true organometallic compounds, but the so-called arene derivatives of the transition metals. A true organochromium compound — $(C_6H_5)_3Cr \cdot 3C_6H_5Li \cdot 2.5(C_2H_5)_2O$ — was only isolated in 1958. This complex was obtained by the reaction between anhydrous chromic chloride and phenyllithium in ether. It is a red-colored solid substance, very unstable in the air. The hydrolysis is accompanied by the formation of benzene; in the absence of oxygen diphenyl is liberated /191/. The reaction between phenylmagnesium bromide and chromic chloride in tetrahydrofuran (THF) at −20°C yielded the tetrahydrofuranate

$(C_6H_5)_3Cr \cdot 3THF$, which is readily hydrolyzed to the green $[Cr(H_2O)_6]^{3+}$ and reacts with corrosive sublimate to give a quantitative yield of phenylmercuric chloride /192, 193/:

$$(C_6H_5)_3Cr \cdot THF + 3HgCl_2 \xrightarrow[-3\,THF]{} 3C_6H_5HgCl + CrCl_3$$

A number of substituted phenyl derivatives of chromium were recently obtained; they include the fine crystalline di-o-anisylchromium /194/, the relatively stable crystalline tri(ω-N-diethylamino)tolylchromium /195/, dimesitylchromium /196/, etc. Of the other organic derivatives of Group VI metals the only ones known are mixed π-cyclopentadienylcarbonyl compounds of molybdenum and tungsten, with a δ-bond between the metal and the radical. They may be prepared by the reaction between the metal hydride and diazomethane /197/:

$$\pi\text{-}C_5H_5Mo(CO)_3H + CH_2N_2 \longrightarrow \pi\text{-}C_5H_5Mo(CO)_3CH_3 + N_2$$

or by the reaction between the sodium salt of metal π-cyclopentadienyltricarbonyl and alkyl halides /198/:

$$\pi\text{-}C_5H_5M(CO)_3Na + RX \longrightarrow \pi\text{-}C_5H_5M(CO)_3R + NaX$$

where if M = Mo, R is CH_3, C_2H_5, or iso-C_3H_7; if M = W, R is CH_3 or C_2H_5.

These compounds are relatively stable yellow crystalline compounds with m.p. between 80 and 150°C. The methyl compound of tungsten is more stable than the molybdenum derivative. All these substances are insoluble in water and slowly decompose under the action of acids and alkalis.

A phenyl derivative of tungsten is known in the form of the complex $(C_6H_5)_4W \cdot 2C_6H_5Li \cdot 3(C_2H_5)_2O$. It is a black crystalline substance, readily soluble in benzene, which easily catches fire in the air /199, 200/.

Of the alkyl derivatives of Group VII metals, methyl compounds of manganese are the only ones known. They may be prepared by the reaction between methyllithium and manganous iodide /188/. Methylmanganese iodide is a brown-colored oil, which is insoluble in ether. Dimethylmanganese has been prepared; it is a yellow powder which spontaneously catches fire and explodes on impact. The reaction with excess CH_3Li yields $[(CH_3)_3Mn]Li$, which is insoluble in ether. Other complexes containing organic manganese derivatives were subsequently prepared: $[(C_2H_5)_3Mn]Li$, $[(C_4H_9)_3Mn]Li$, $[(C_6H_5)_3Mn]Li$, etc. /202/.

Diphenylmanganese is prepared from phenylmagnesium bromide and manganese chloride or from phenyllithium and manganese iodide at −50°C. Tetrahydrofuran is used as solvent /203/. Stable alkyl and phenyl derivatives of manganese and rhenium, in which the metal is bound both to the hydrocarbon radical and to carbonyl groups, are also known /204/. For example, pentacarbonylmethylmanganese forms colorless crystals, m.p. 95°C, which are stable in the air. Organic derivatives of rhenium have been obtained in the same manner /205/.

Platinum is the only Group VIII metal which forms stable alkyl derivatives. Methyl derivatives of Pt(IV) have been studied in detail — trimethylplatinum iodide, tetramethylplatinum, etc. They are prepared by reacting methylmagnesium iodide with $PtCl_4$ and its complexes /160, 208/. Alkyl and aryl derivatives of other metals in Group VIII were obtained only as complexes which contains ligands capable of stabilizing the σ-bond between carbon and metal (e.g., C_5H_5, CO, substituted phosphines, etc.) /160, 206/. Very probably, organic derivatives of Group VIII metals are formed as intermediate compounds during the reactions between the salts of these metals and alkyl or aryl compounds of lithium, magnesium, zinc, etc. /181/. This is indicated by the change in the color of the solutions and by the adsorption of hydrogen. This is accompanied by dimerization or disproportionation of the organic radicals of the reacting organometallic compounds. However, these compounds undergo rapid transformations, up to the full decomposition of the product. Mixed π-cyclopentadienyl and carbonyl compounds of iron of the type π-$C_5H_5M(CO)_nR$ are known, which have a σ-bond with methyl, ethyl, phenyl and cyclopentadienyl groups. They are colored, diamagnetic products with sharp melting points /197/, soluble in organic solvents, which easily sublime in high vacuum at 25 — 50°C and have a characteristic odor. All these compounds are unstable in organic solvents; they are insoluble in water, but are decomposed by acids and bases.

The known alkyl derivatives of cobalt include the unstable $CH_3Co(CO)_4$, which is readily rearranged to acetylcobalttricarbonyl $CH_3COCo(CO)_3$ /207/. Tetrahydrofuranates of diphenylcobalt and dimesitylcobalt /196, 203/ have also been isolated.

The isolated and identified nickel derivatives include the complexes R_2Ni with triethylphosphine or triphenylphosphine $(R'_3P)_2NiR_2$ and $(R'_3P)_2Ni(R)X$. They are prepared by the reaction between the corresponding dihalide and the organomagnesium compound in liquid ammonia /209/. They are bright-colored products, mostly yellow-brown. Compounds with phenyl, o-tolyl and other aryl radicals have been isolated. They are diamagnetic, symmetrical and have a small dipole moment. Their chemical reactions indicate that the metal-radical bonds are mobile. σ-Alkyl complexes of nickel of the composition π-$C_5H_5NiP(C_6H_5)_3R$, where $R = CH_3$, C_2H_5, C_6H_5, etc., were recently prepared by the reaction between RMg and cyclopentadienyltriphenylphosphinenickel chloride. They are thermally stable, but are readily decomposed in solution if in contact with the atmosphere /232/.

Bibliography

1. Wierl, R. — Ann. Phys., **5**(8):521. 1931.
2. Fischer, W. and A. Taurinsch. — Z. anorg. allg. Chem., **205**:309. 1932.
3. Fowler, A. — Proc. Roy. Soc., **79A**:509. 1907.
4. Hukumoto, J. — Sci. Rep. Tohoku Imp. Univ., **22**:868. 1933.
5. Daure, P. — Ann. Phys., **12**:357. 1929; Compt. rend., **187**:940. 1929.
6. Luchinskii, G. P. — ZhOKh, **7**:207. 1937.
7. Arii, K. — Sci. Rep. Tohoku Imp. Univ., **22**:182. 1933.
8. Stillvell, Ch. — Uspekhi Khimii, **6**:1174. 1937.
9. Sharma, P. — Bull. Acad. Sci. Allahabad, **3**:87. 1933.
10. Luchinskii, G. P. — ZhFKh, **6**:607. 1935.
11. Holtje, R. — Z. anorg. allg. Chem., **209**:241. 1932.
12. Regnault, R. — Ann. chim. phys., **1**:129. 1841.
13. Beeguerel, B. — Ann. chim. phys., **5**:82. 1877.
14. Luchinskii, G. P. Chetyrekhkhloristyi titan (Titanium Tetrachloride), p. 24. — Oborongiz. 1939.

15. Latimer, W. — J. Am. Chem. Soc., **44**:90. 1922.
16. Gutbier, A. — Z. angew. Chem., **28**(1):124. 1915.
17. Schumb, W. and R. Sundstrom. — J. Am. Chem. Soc., **55**:596. 1933.
18. Parravano, N. and G. Mazzetti. — Rec. trav. chim., **42**:821. 1923.
19. Billy, M. — Ann. chim., **16**:5. 1921.
20. Hunter, H. — J. Am. Chem. Soc., **32**:330. 1910.
21. Ruff, O. and F. Neumann. — Z. anorg. allg. Chem., **128**:81. 1923.
22. Hock, L. and W. Knauf. — Z. anorg. allg. Chem., **228**:193. 1936.
23. Billy, M. et al. — Compt. rend., **200**:1765. 1935.
24. Luchinskii, G. P. Chetyrekhkhloristyi titan (Titanium Tetrachloride), p. 35. — Oborongiz. 1939.
25. Kowalewsky, H. — Z. anorg. Chem., **25**:185. 1900.
26. Schwarz, R. and H. Richter. — Ber., **62**:31. 1929.
27. Biltz, W. and E. Kennecke. — Z. anorg. allg. Chem., **147**:171. 1925.
28. Ralston, A. and J. Wilkinson. — J. Am. Chem. Soc., **50**:258. 1928.
29. Rosenheim, A. and O. Schütte. — Z. anorg. Chem., **24**:238. 1901.
30. Stähler, H. — Ber., **38**:2619. 1905.
31. Shretien, A. and G. Varga. — Compt. rend., **201**:558. 1935.
32. Bourion, B. — Ann. chim. phys., **21**:49. 1910.
33. Ruff, O. — Ber., **37**:4513. 1904.
34. Rose, H. — Lieb. Ann., **24**:141. 1830.
35. Jeunings, J. et al. — J. Am. Chem. Soc., **54**:637. 1936.
36. Puxeddu, E. — Gazz. chim. ital., **59**:160. 1929.
37. Rosenheim, A. et al. — Ber., **36**:1833. 1903.
38. Dermer, O. and W. Fernelius. — Z. anorg. allg. Chem., **221**:83. 1934.
39. Schmitz-Dumont, O. and E. Motzkus. — Ber., **62**:466. 1929.
40. Rosenheim, A. and O. Schütte. — Z. anorg. Chem., **26**:239. 1901.
41. Badische Anilin und Soda-Fabrik A.G., British Patent 840619. 1960.
42. Farbwerke Hoechst A.G., US Patent 2951085. 1960.
43. Farbwerke Hoechst A.G., Belgian Patent 553477. 1957.
44. Schumb, H. and R. Sundström. — J. Am. Chem. Soc., **55**:596. 1933.
45. Natta, G. and I. Pasquon. — Kinetika i Kataliz, **3**(6):805. 1962.
46. Natta, G. et al. — J. Polymer Sci., **51**:399. 1961.
47. Korotkov, A.A. and Li Tsung-ch'ang. — Vysokomolekulyarnye Soedineniya, **3**(5):691. 1961.
48. Wedekind, E. and P. Hausknecht. — Ber., **46**:3763. 1913.
49. Dijkgraaf, C. and J. Rousseau. — Spectrochim. acta, **23A**(5):1267. 1967.
50. Meyer, F. et al. — Ber., **56**:1908. 1923.
51. Pfordten, O. Von. — Lieb. Ann., **237**:201. 1887.
52. Friedel, C. and J. Guerin. — Compt. rend., **81**:889. 1923.
53. Pigeard, J. — J. Am. Chem. Soc., **48**:2295. 1926.
54. Stähler, A. and F. Wirthwein. — Ber., **38**:2619. 1905.
55. Georges, H. and A. Stähler. — Ber., **42**:3200. 1909.
56. Boenziser, N. and R. Rundle. — Acta crystallogr. Cambridge, **1**:274. 1948.
57. Nisel'son, L.A. and K.V. Tret'yakova. — ZhNKh, **12**(4):857. 1967.
58. Friedel, C. and J. Guerin. — Compt. rend., **81**:889. 1923.

59. Schumb, H. and R. Sundström. — J. Am. Chem. Soc., **55**:596. 1933.
60. Klemm, W. and L. Grimm. — Z. anorg. allg. Chem., **249**:209. 1942.
61. Spiridonov, V. P. et al. — Zhurnal Strukturnoi Khimii, **3**:329. 1962.
62. Nekrasov, B.V. Kurs obshchei khimii (A Course in General Chemistry). — Goskhimizdat. 1962.
63. Berdonosov, S.S. et al. — ZhNKh, **7**:1465. 1962.
64. Tsirel'nikov, V.I. et al. — Doklady AN SSSR, **146**:122. 1962.
65. Razuvaev, G.A., L.I. Bobinova, and V.S. Etlis. — Doklady AN SSSR, **122**:618. 1958.
66. Deluzarche, A. — Comptes rendus, **239**:1489. 1954.
67. Takatini, T. — Bull. Chem. Soc. Japan, **7**:705. 1957;
68. Yamomoto, A. and S. Kambara. — J. Am. Chem. Soc., **79**:4344. 1957.
69. Kriegsman, H. and K. Licht. — Z. Electrochem., **62**(10):1163. 1958.
70. Pande, K. and R. Mehrotra. — Chem. Ind. (London), No. 37:1198. 1958.
71. Chakravarti, B. — Naturwiss., **45**:286. 1958.
72. Suvorov, A.A. and S.S. Spasskii. — Uspekhi Khimii, 28(11):1277. 1959.
73. Andreev, Yu.N. and V.A. Nikol'skii. — Sbornik Statei po Obshchei Khimii, **11**:1428. 1953.
74. French Patent 818570. 1937; C.A., **32**:2545. 1938.
75. Nogina, O.V., R.Kh. Freidlina, and A.N. Nesmeyanov. — Izvestiya AN SSSR, OKhN, p. 327. 1950.
76. Cullinane, N. and S. Chard. — Nature, **164**:710. 1949.
77. Haslam, J. — US Patent 2684972; C.A., **49**:10999. 1955.
78. Herman, D. — US Patent 2654770; C.A., **48**:1370. 1954.
79. Shiihara, I., W.T. Schwartz, Jr., and H.W. Post. — Uspekhi Khimii, **34**(1):44. 1965.
80. Lowe, W. — US Patent 2795553. 1957; C.A., **51**:14255. 1957.
81. Nesmeyanov, A.N., E.M. Brainina, and R.Kh. Freidlina. — Doklady AN SSSR, **94**:249. 1954.
82. Cox, E. — J. Chem. Soc., p. 731. 1935.
83. Yamasaki, K. and K. Sone. — Nature, **166**:998. 1950.
84. Uitert, V. and C. Fernelius. — J. Am. Chem. Soc., **75**:2736. 1953.
85. Gach, H. — Monatsh., **21**:116. 1900.
86. Urbain, L. and H. Debierne. — Compt. rend., **129**:303. 1899.
87. British Dyestuffs, Corp., British Patent 289493. 1927.
88. Biltz, W. — Ann., **331**:348. 1904.
89. Cortledse, G. — J. Ann. Chem. Soc., **73**:4416. 1951.
90. Ctacca, G. — Gazz. chim. ital., **67**:316. 1937.
91. Menzies, U. — J. Chem. Soc., p. 1755. 1934.
92. Hirst, H. and G. Bruxelles. — Ind. Eng. Chem., **48**:1325. 1956.
93. Blackmore, N. — Chem. Products, **23**:367. 1960.
94. Phillips Petroleum Co., Belgian Patent 552578. 1956.
95. Schäfer, H. — Z. anorg. allg. Chem., **353**(5—6):281. 1967.
96. Pflugmacher, A. et al. — Naturwiss., **45**:490. 1958.
97. Dilthey, W. — Ann., **344**:300. 1906.
98. Pande, K. and R. Mehrotra. — Chem. Ind. (London), No. 37:1198. 1958.
99. Chakravarti, B. — Naturwiss., **45**:286. 1958.

100. Larsen, E. et al. — J. Am. Chem. Soc., **75**:5107. 1953.
101. Biltz, W. and J. A. Clinch. — Z. anorg. Chem., **40**:221. 1904.
102. Shigorin, D. N. — ZhFKh, **27**:554. 1953.
103. Anderson, A. — Chem. Week, **87**:70. 1960; Chem. Week, **86**:101. 1960; Chem. Eng. News, **38**:63. 1960.
104. Harwood, J. Industrial Applications of Organometallic Compounds. — London, Chapman and Hall, p. 366. 1963.
105. Jones, M. — J. Am. Chem. Soc., **81**:3188. 1959.
106. Fernelius, C. et al. — Inorg. Synthesis, **6**:147. 1960.
107. BASF, GFR Patent 1039056. 1958.
108. Hautzsch, A. and C. Desch. — Ann., **323**:1. 1902.
109. BASF, GFR Patent 1044801. 1958.
110. See /104/, p. 374.
111. BASF, GFR Patent 1076121. 1960.
112. Gach, H. — Monatsh., **21**:116. 1900.
113. Krause, E. and A. von Grosse. Die Chemie der metallorganischen Verbindungen. — Berlin. 1937.
114. Jaffe, H. and G. Doak. — J. Chem. Phys., **21**:196. 1953.
115. Kealy, T. and P. Pauson. — Nature, **168**:1039. 1951.
116. Nesmeyanov, A. N. — Uspekhi Khimii, **14**:261. 1945.
117. Nesmeyanov, A. N. — Uspekhi Khimii, **28**:1163. 1959.
118. Razuvaev, G. A. and V. N. Latyaeva. — Uspekhi Khimii, **4**:585. 1965.
119. Zeiss, H. (editor). Organometallic Chemistry. — New York, Reinhold. 1960.
120. Rochow, E. G., D. T. Hurd, and R. N. Lewis. The Chemistry of Organometallic Compounds. — New York, Wiley. 1957.
121. Hein, F. — Ber., **52**:195. 1919.
122. See /119/.
123. Zeiss, H. et al. — Angew. Chem., **67**:282. 1955.
124. Fischer, E. and R. Jira. — Z. Naturforsch., **8b**:1. 1953.
125. Wilkinson, G. et al. — J. Am. Chem. Soc., **75**:1011. 1953.
126. Wilkinson, G. and J. Birmingham. — J. Am. Chem. Soc., **76**:4281. 1954.
127. Thomas, J. — US Patent 2680758. 1954; C.A., **49**:4725. 1955.
128. Wilkinson, G. and F. Cotton. — Chem. a. Ind., p. 307. 1954.
129. Cotton, F. and G. Wilkinson. — Z. Naturforsch., **9b**:417. 1954.
130. Riemschneider, R. and D. Helm. — Ber., **89**:155. 1956.
131. Wilkinson, G. et al. — J. Am. Chem. Soc., **76**:1970. 1954.
132. Wilkinson, G. — J. Am. Chem. Soc., **74**:6148. 1953.
133. Wilkinson, G. — J. Am. Chem. Soc., **74**:6146. 1953.
134. Cotton, F. et al. — J. Am. Chem. Soc., **75**:3586. 1954.
135. Wilkinson, G. et al. — J. Inorg. Nucl. Chem., **2**:95. 1956.
136. Wilkinson, G. and J. Birmingham. — J. Am. Chem. Soc., **76**:6210. 1954.
137. Birmingham, J. and G. Wilkinson. — J. Am. Chem. Soc., **78**:42. 1956.
138. Fischer, E. and U. Piesbergen. — Z. Naturforsch., **11b**:758. 1958.
139. Wilkinson, G. et al. — J. Am. Chem. Soc., **74**:2125. 1952.
140. Dunitz, J. and L. Orgel. — Nature, **171**:121. 1953.
141. Dunitz, J. and L. Orgel. — J. Chem. Phys., **23**:954. 1955.

142. Moffitt, W. — J. Am. Chem. Soc., **76**:3386. 1954.
143. Jaffe, H. — J. Chem. Phys., **21**:156. 1953.
144. Yamazaki, M. — J. Chem. Phys., **24**:1260. 1956.
145. Schwab, G. et al. — Naturwiss., **41**:228. 1954.
146. Eiland, P. and R. Repinsky. — J. Am. Chem. Soc., **74**:4971. 1952.
147. Struchkov, Yu. T. — ZhOKh, **27**:2039. 1957.
148. Dyatkina, M. E. — Uspekhi Khimii, **27**(1):57. 1958.
149. Nesmeyanov, A. N. and E. G. Perevalova. — Uspekhi Khimii, **27**(1):3. 1958.
150. Summers, L. and R. Uloth. — J. Am. Chem. Soc., **76**:2278. 1954.
151. Summers, L. et al. — J. Am. Chem. Soc., **77**:3604. 1955.
152. Fischer, E. and G. Wilkinson. — J. Inorg. Nucl. Chem., **2**:149. 1956.
153. Birmingham, J. et al. — Naturwiss., **42**:96. 1955.
154. Cotton, F. et al. — J. Inorg. Nucl. Chem., **1**:175. 1955.
155. Engelmann, F. — Z. Naturforsch., **8b**:775. 1953.
156. Fischer, E. and R. Jira. — Z. Naturforsch., **9b**:618. 1954.
157. Fischer, E. — Z. Naturforsch., **9b**:619. 1954.
158. Weiss, E. — Z. anorg. allg. Chem., 287:236. 1956.
159. Fischer, E. and W. Pfab. — Z. anorg. allg. Chem., **274**:316. 1953.
160. Zeiss, H. (editor). Organometallic Chemistry. — New York, Reinhold. 1960. Kritskaya, I. I. — Uspekhi Khimii, **35**:393. 1966.
161. Beerman, C. — Angew. Chem., **71**:195. 1959.
162. Pittser, V. M. — ZhOKh, **8**:1298. 1938.
163. Beerman, C. and H. Bestian. — Angew. Chem., **71**:618. 1959.
164. GFR Patent 1100022. 1961.
165. Clauss, K. and C. Beerman. — GFR Patent 1046048. 1959.
166. Herman, D. and W. Nelson. — J. Am. Chem. Soc., **75**:3877. 1953.
167. Vries, H. de. — Rec. trav. chim. Pays-Bas, **80**:866. 1961.
168. Razuvaev, G. A. et al. — Doklady AN SSSR, **156**:1121. 1964.
169. Bawn, C. and J. Gladstone. — Proc. Chem. Soc., p. 227. 1959..
170. Berthold, H. and G. Groh. — Z. anorg. allg. Chem., **319**:230. 1963.
171. Razuvaev, G. A. et al. — Doklady AN SSSR, **156**:1121. 1964.
172. Razuvaev, G. A. et al. — ZhOKh, **32**:1354. 1962.
173. Clauss, K. and H. Bestian. — Ann. Chem., **654**:8. 1962.
174. Razuvaev, G. A. et al. — ZhOKh, **31**:2667. 1961.
175. Walter, A. — US Patent 3027392. 1963.
176. Peters, W. — Chem. Ber., **41**:3173. 1958.
177. Gilman, H. and R. Jones. — J. Org. Chem., **10**:505. 1945.
178. Berthold, H. and G. Groh. — Angew. Chem., **75**:576. 1963.
179. Halloway, H. — Chem. a. Ind., p. 214. 1962.
180. Kawasaki Yoshikane, et al. — Bull. Chem. Soc. Japan, **40**(7): 1562. 1967.
181. Cotton, F. — Chem. Rev., **55**:551. 1955.
182. Kurras, E. — Monatsber. Dtsch. Acad. Wiss. Berlin, **2**:109. 1960.
183. Carrik, W. et al. — J. Am. Chem. Soc., **82**:5319. 1960.
184. GFR Patent 1085156. 1960.
185. Liefde, H. et al. — Chem. Ind., p. 119. 1960.
186. Afanas'ev, V. N. — Z. anorg. allg. Chem., **245**:381. 1941.
187. Sarry, B. and V. Dobrusskin. — Angew. Chem., **74**:509. 1962.
188. Beerman, C. and K. Clauss. — Angew. Chem., **71**:627. 1959.

189. Piper, T. and G. Wilkinson. — Chem. Ind., **41**:1296. 1955.
190. Hein, F. and W. Melms-Bode. — Z. anorg. allg. Chem., **51**:503. 1938.
191. Hein, F. and R. Weiss. — Z. anorg. allg. Chem., **295**:145. 1958.
192. Zeiss, H. and W. Herwig. — J. Am. Chem. Soc., **79**:6561. 1957.
193. Zeiss, H. and W. Herwig. — J. Am. Chem. Soc., **81**:4798. 1959.
194. Hein, F. and D. Tille. — Monatsber. Dtsch. Akad. Wiss. Berlin, **4**:414. 1962.
195. Bahr, G. and H. Zohm. — Angew. Chem., **75**:110. 1963.
196. Tsutsui, M. and H. Zeiss. — J. Am. Chem. Soc., **82**:6255. 1960.
197. Piper, T. and G. Wilkinson. — J. Inorg. Nucl. Chem., **3**:104. 1956/1957.
198. Sinn, H. and F. Patat. — Angew. Chem., **75**:805. 1963.
199. Funk, H. and W. Hanke. — Angew. Chem., **71**:408. 1959.
200. Sarry, B. and M. Dettke. — Angew. Chem., **75**:1022. 1963.
201. Juvinall, G. L. — J. Am. Chem. Soc., **86**:4202. 1964.
202. Riemschneider, R. et al. — Naturforsch., **15b**:547. 1960.
203. Tsutsui, M. and H. Zeiss. — J. Am. Chem. Soc., **83**:825. 1961.
204. Clossen, R. et al. — J. Org. Chem., **22**:598. 1957.
205. Hieber, W. and G. Braun. — Naturforsch., **14b**:132. 1959.
206. Hallam, B. and P. Pauson. — Chem. Ind., **23**:653. 1955.
207. Hieber, W. et al. — Naturforsch., **13b**:192. 1958.
208. Gilman, H. and M. Lichtenwalter. — J. Am. Chem. Soc., **60**:3085. 1938; J. Am. Chem. Soc., **75**:2063. 1953.
209. Kharasch, M. and O. Reinmuth. Grignard Reactions of Nonmetallic Substances. — New York, Prentice-Hall. 1954.
210. Reznichenko, V. A. and V. P. Solomakha. — In: "Titan i ego splavy," Vol. 4, Izd. AN SSSR. 1960.
211. Reznichenko, V. A. and V. P. Solomakha. — In: "Titan i ego splavy," Vol. 4, Izd. AN SSSR. 1961.
212. Reznichenko, V. A. and V. P. Solomakha. — In: "Titan i ego splavy," Vol. 8:89, Izd. AN SSSR. 1962.
213. Galitskii, N. V. and S. V. Shadskii. — In: "Titan i ego splavy," Vol. 8:140, Izd. AN SSSR. 1962.
214. Furman, A. A. and V. B. Lavrova. — Khimicheskaya Promyshlennost', **5**:42. 1966.
215. Krieve, W. and D. Mason. — J. Chem. Phys., **25**:3. 1956.
216. Meyer, F. et al. — Ber., **56**:1908. 1923.
217. Ingraham, T., K. Downes, and P. Marier. — Canad. J. Chem., **35**(8):850. 1957.
218. Chem. Week, **80**:65. 1957.
219. Razuvaev, G. A. et al. — Doklady AN SSSR, 173(6):353. 1967.
220. Gernes, D. and R. Edwin. — British Patent 757873. 1954.
221. Yablokov, Yu. S. and V. M. Lozovatskii. — Soviet Patent 125244. 1959.
222. Yablokov, Yu. S. and V. M. Lozovatskii. — In: "Titan i ego splavy," **8**:135, Izd. AN SSSR. 1962.
223. Ruherwein, R. and G. Skinner. — US Patent 2720445. 1955.
224. Ruherwein, R. and G. Skinner. — US Patent 2734479. 1956.
225. Kutsev, V. S. — ZhFKh, **31**(8):1866. 1957.
226. Ryabov, V. A. and Z. N. Zviadadze. — Trudy Instituta Metallurgii AN SSSR im. Baikova, **1**:85. 1957.

227. Seelbach, C. — US Patent 2925392. 1960.
228. Furman, A.A. and V.B. Lavrova. — Soviet Patent 142639. 1960; Byulleten' Izobretenii, No. 22. 1960.
229. Korotkov, A.A, and Li Tsung-ch'ang. — Vysokomolekulyarnye Soedineniya, **3**(4):686. 1961.
230. Pfeffer, G. Dissertation Technische Hochschule, Stuttgart. 1954.
231. Meyer, H., M. Janssen, and G. Kerk. — Rec. trav. chim. Pays-Bas, **80**:831. 1961.
232. Yamazaki, H. and N. Hagihara. — Bull. Chem. Soc. Japan, **37**:907. 1964.

Chapter III

THE STUDY OF COMPLEX ORGANOMETALLIC COMPOUNDS

Complex organometallic catalysts are prepared by reacting various compounds of transition metals in Groups IV — VIII with organic compounds of metals in Groups I — III /1/. Metal alkyls which can be used as constituents of the catalyst complex include aluminum, beryllium, zinc, magnesium, lithium, sodium /2/ and cadmium /3, 4/ alkyls. Mixed alkyls of alkali metals with aluminum, gallium, indium and thallium can also be employed /2/. The second component of complex organometallic systems is most often a halide or an alkoxide of titanium, chromium, vanadium /2/, zirconium, thorium, uranium, iron /3/, etc.

However, not all of these compounds and not all of their combinations will give the same effect in the polymerization of olefins or in other processes. The choice of the components of the catalyst complex, their mutual proportion, their physical state and the experimental conditions have a major effect on the properties of the polymer product, on the reaction rate, etc.

Catalyst complexes most often used in polymerization processes have an alkylaluminum compound as the organometallic constituent and titanium or vanadium halide or alkoxide in the capacity of the salt of the transition metal. The different alkylaluminum compounds such as triethylaluminum, diethylaluminum chloride, diethylaluminum ethoxide, triisobutylaluminum, diisobutylaluminum chloride, etc., do not give the same effect when combined with a given salt of a transition metal. The main reason for it is the fact that the formation of an active catalyst complex involves the alkylation of the transition metal salt (such as $TiCl_4$) and its subsequent reduction. These properties differ for the different organoaluminum compounds. It has been shown that the reducing power decreases in the sequence $R_2AlH > AlR_3 > R_2AlX > RAlX_2$, where X is a halogen or an alkoxy group, and the reducing power of R_2AlH is greater than its alkylation capacity. It is also known that when trialkylaluminum compounds react with another substance, the different Al — C bonds are not equivalent: the first bond is the most reactive, while the third bond is relatively inert.

The mode of preparation of the catalyst complex largely affects its activity. At least five different modes of preparation must be distinguished:

1. Solutions of the alkylaluminum compound and $TiCl_4$ (or another compound used instead of $TiCl_4$) are separately introduced into the reaction mixture, the formation of the active catalyst complex proceeding in the presence of the monomer and of the diluent.

2. The alkylaluminum compound and the $TiCl_4$ are fed into the reactor as gases, in a stream of ethylene; the formation of the catalyst then takes place in the presence of ethylene in the gas phase.

3. The catalyst is prepared in the absence of the monomer by mixing solutions of components under definite conditions, at controlled temperature and rate of stirring. The extent of reduction of $TiCl_4$ is controlled analytically /7/ or by holding for a definite period of time /8/, after which the catalyst is fed into the reactor.

4. The catalyst is mixed together first, the precipitate is isolated and a given amount of the alkylaluminum compound is added. This technique is similar to that proposed by Natta, in which $TiCl_3$ is used in lieu of $TiCl_4$. The method yields long-lived catalysts; also, the polymerization process is more easily monitored.

5. The catalyst is prepared in the presence or in the absence of the monomer, but the effect of the catalyst is modified during the polymerization by adding one of the constituents of the catalyst, or substances such as amines, alcohols, thiophenols /7/, oxygen /9/, hydrochloric acid /10/, etc.

The simplest method — mixing of constituents under definite experimental conditions in the absence of the monomer — has also proved the most reliable. The preparation of catalyst complexes from other constituents such as alkyltin compounds, halogenated derivatives of vanadium and aluminum is also effected in this manner.

In order to clarify the presumed reaction mechanism of polymerization processes, we shall discuss in detail the reactions which take place between the constituents of the catalyst complexes.

1. REACTIONS BETWEEN ALKYLALUMINUM COMPOUNDS AND TITANIUM TETRACHLORIDE

The reaction between triethylaluminum and titanium tetrachloride was repeatedly studied /11 — 19/ as a typical example of the reactions involved in the catalytic effect of the complex. These compounds react with the evolution of gases; the precipitate which separates out at the same time no longer contains tetravalent titanium, but rather titanium which has been reduced, to a greater or lesser extent, to trivalent or even to bivalent titanium. The formation of the precipitate and the reduction of titanium is due to the alkylation of $TiCl_4$:

$$AlR_3 + TiCl_4 \longrightarrow RTiCl_3 + R_2AlCl$$

with subsequent rupture of the Ti — C bond in the organotitanium compound. The diethylaluminum chloride formed as a result of the reaction, continues to react with $TiCl_4$:

$$R_2AlCl + TiCl_4 \longrightarrow RTiCl_3 + RAlCl_2$$

The titanium trichloride which is formed by decomposition of alkyltitanium trichloride

$$RTiCl_3 \longrightarrow TiCl_3 + R$$

also becomes alkylated with formation of $RTiCl_2$, which is bound as coordination complex /20/. Both the alkylation and the decomposition of $RTiCl_3$ are reactions which take time, and their rates are different for different alkylaluminum compounds. The reaction rates, and thus also the composition of the precipitate at any given moment depends on the reaction temperature, concentrations of initial substances, their mutual ratio, time and stirring rate. Another important factor is the mode of intermixing of the constituents of the complex.

The reaction between triethylaluminum and titanium tetrachloride was first described by Ziegler /21, 22/, who noted that the reaction took place at room temperature, with evolution of gaseous products and separation of a precipitate which contained hydrocarbon groups, chloride, aluminum and titanium.

The elementary composition of the precipitate formed at different ratios between triethylaluminum and $TiCl_4$ was first studied by Natta et al. /11/ who found that, as the molar ratio of the reactants $Al(C_2H_5)_3:TiCl_4$ was increased from 1 to 4, the Cl:Ti ratio in the precipitate decreased from 2.7 to 0.7, whereas the Al:Ti ratio increased from 0.2 to 0.4.

Friedländer and Oita /18/ studied the reaction between triisobutylaluminum and $TiCl_4$ and found that if the reagents are used in equimolar proportions, titanium is reduced to Ti(III); insignificant amounts of Ti(II) were noted in isolated experiments only.

The first quantitative data on the effect of temperature and concentration of the reagents on the reaction note were published by Pozamantir et al. /17/. Figure 3 shows the apparatus employed in the study of the reactions between alkylaluminum compounds and $TiCl_4$ in the absence of solvents. Titanium tetrachloride was introduced into one limb of vessel 1, while the alkylaluminum compound was introduced into the other limb. The reagents were introduced from the metering vessel 2. The vessel was then placed in a thermostat and connected to a gasometric system. Vessel 1 was then cautiously tilted and the reactants were thus brought into contact at room temperature. Since the reagents are extremely reactive, the mixing was conducted very slowly (during 15 — 20 minutes).

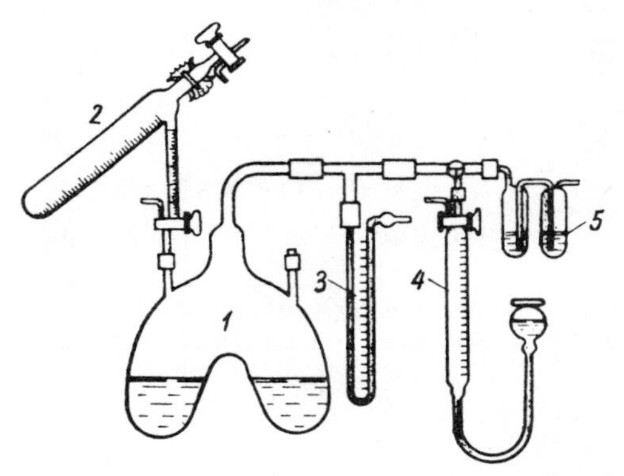

FIGURE 3. Diagram of apparatus used to conduct the reaction between $TiCl_4$ and $AlR_{3-n}X_n$ in the absence of solvent:

1 — mixing vessel; 2 — metering vessel; 3 — manometer; 4 — gas buret; 5 — hydraulic seal.

Experimental results showed that the reaction rate between $C_2H_5AlCl_2$ and $TiCl_4$ increases with increasing temperature. Not more than 80% of

Ti(IV) present could be reduced. That $C_2H_5AlCl_2$ was not fully consumed was confirmed by the fact that the precipitate which separated out in the presence of a large excess of $TiCl_4$ evolved ethane on being decomposed with water. However, if $C_2H_5AlCl_2$ was present in double the stoichiometric amount, all titanium was converted to Ti(III). The effect of the temperature and the molar ratio of the reagents on the rate of the reaction between $(C_2H_5)_2AlCl$ and $TiCl_4$ was similar, except that at room temperature $TiCl_4$ was more speedily reduced than by $C_2H_5AlCl_2$, while at 40°C the rates of the two reactions were practically the same. The dependence of the rate of formation of Ti(III) on the temperature and the molar ratio was similar for the reaction between $Al(C_2H_5)_3$ and $TiCl_4$ as well, but the reaction rate at room temperature was much faster: thus, only 2 minutes after the reagents were mixed, the extent of formation of Ti(III) was about 33%. Simon et al. /13, 19, 23/ carried out the most detailed study of the effect of the molar ratio of the reactants and temperature on the composition of the precipitates and on the amount and the composition of the gases evolved. In the range of molar ratios $Al(C_2H_5)_3:TiCl_4$ between 1.5 and 5.0 the precipitate contained two Ti atoms per one Al atom, and only when the ratio was less than 1.5, did the amount of aluminum in the precipitate slowly decrease. When these ratios were employed, the content of chlorine varied from 70 to 30%, while the amount of the ethyl groups in the precipitate increased from 10 to 40%. On the strength of these data, the formula $AlTi_2(C_2H_5)_nCl_{7-n}$, where $n = 2, 3, 4$, was proposed for the precipitate formed by the reaction between triethylaluminum and titanium tetrachloride. The authors deduced the following reaction mechanism and proposed the following chemical reactions as taking place at the different values of the molar ratio:

Reaction	Molar ratio $AlR_3 : TiCl_4$
$3Al(C_2H_5)_3 + 2TiCl_4 \longrightarrow AlTi_2Cl_5(C_2H_5)_2 + Al(C_2H_5)_2Cl + Al(C_2H_5)Cl_2 + 4C_2H_5$	1.5
$4Al(C_2H_5)_3 + 2TiCl_4 \longrightarrow AlTi_2Cl_5(C_2H_5)_2 + 3Al(C_2H_5)_2Cl + 4C_2H_5$	2.0
$3Al(C_2H_5)_3 + 2TiCl_4 \longrightarrow AlTi_2Cl_4(C_2H_5)_3 + 2Al(C_2H_5)Cl_2 + 4C_2H_5$	1.5
$4Al(C_2H_5)_3 + 2TiCl_4 \longrightarrow AlTi_2Cl_4(C_2H_5)_3 + Al(C_2H_5)Cl_2 + 2Al(C_2H_5)_2Cl + 4C_2H_5$	2.0
$5Al(C_2H_5)_3 + 2TiCl_4 \longrightarrow AlTi_2Cl_4(C_2H_5)_3 + 4Al(C_2H_5)_2Cl + 4C_2H_5$	2.5
$4Al(C_2H_5)_3 + 2TiCl_4 \longrightarrow AlTi_2Cl_3(C_2H_5)_4 + 2Al(C_2H_5)Cl_2 + Al(C_2H_5)_2Cl + 4C_2H_5$	2.0
$5Al(C_2H_5)_3 + 2TiCl_4 \longrightarrow AlTi_2Cl_3(C_2H_5)_4 + Al(C_2H_5)Cl_2 + 3Al(C_2H_5)_2Cl + 4C_2H_5$	2.5
$6Al(C_2H_5)_3 + 2TiCl_4 \longrightarrow AlTi_2Cl_3(C_2H_5)_4 + 5Al(C_2H_5)_2Cl + 4C_2H_5$	3.0

If the molar ratio between triethylaluminum and $TiCl_4$ is less than 1.5, the precipitate contains a different amount of aluminum, and the course of the reaction becomes much more complex. Thus, it is maintained by the authors that if the molar ratio $Al(C_2H_5)_3:TiCl_4$ is between 0.75 and 1.25, the following reactions may take place, in conformity with the empirical formula of the precipitate $AlTi_4Cl_{12}(C_2H_5)_2$:

Reaction	Molar ratio $Al(C_2H_5)_3 : TiCl_4$
$3Al(C_2H_5)_3 + 4TiCl_4 \longrightarrow AlTi_4Cl_{12}(C_2H_5)_2 + 2Al(C_2H_5)Cl_2 + 5C_2H_5$	0.75
$4Al(C_2H_5)_3 + 4TiCl_4 \longrightarrow AlTi_4Cl_{12}(C_2H_5)_2 + 2Al(C_2H_5)_2Cl + Al(C_2H_5)Cl_2 + 5C_2H_5$	1.0
$5Al(C_2H_5)_3 + 4TiCl_4 \longrightarrow AlTi_4Cl_{12}(C_2H_5)_2 + 4Al(C_2H_5)_2Cl + 5C_2H_5$	1.25

With the increase in the $Al(C_2H_5)_3:TiCl_4$ ratio, the amount of the gases evolved increases. If $Al_2(C_2H_5)_3Cl_3$ and $Al(C_2H_5)_2Cl$ are used instead of triethylaluminum, the reaction with $TiCl_4$ is accompanied by a much less intense gas evolution. It is interesting to note in this connection that the reactions of $TiCl_4$ with $Al(C_2H_5)_2Cl$ and with $Al_2(C_2H_5)_3Cl_3$ result in the evolution of approximately equal volumes of gas, even though the former compound contains one ethyl group less. The explanation given by the authors for this effect is that the liberated gases consist mainly of ethane end ethylene: the amount of the former is practically equal in both cases, while the latter polymerizes to yield liquid products.

A study of the valency of titanium in the precipitate showed that the volume of the gas evolved in the reaction between the alkylaluminum compounds and $TiCl_4$, was proportional to the decrease in the valency of titanium. It was shown that each liberated ethyl group corresponds to a decrease of one unit in the valency of titanium. The average valency of titanium according to the equation decreased by 1.25 with respect to the initial $TiCl_4$, which means that one-quarter of the titanium was bivalent, in addition to the trivalent titanium.

The reaction between triisobutylaluminum and titanium tetrachloride shows similar relationships. It has been shown /31/ that after the reactants are intermixed at room temperature, a dark brown precipitate separates out; its composition as function of the ratio between the reactants is shown in Table 9.

TABLE 9. Analysis of the complex $Al(iso\text{-}C_4H_9)_3 - TiCl_4$

$Al(iso\text{-}C_4H_9)_3$: : $TiCl_4$ ratio	Ti, %		Cl, %	Al, %	R, %	Atomic ratios			
	reduced	total				Al : Ti	Cl : Ti	R : Ti	R : Al
0.5	5.4	30.6	49.0	7.86	12.4	0.46	2.16	0.34	0.75
0.5	3.2	23.0	39.1	—	—	—	2.30	—	—
1.0	—	26.5	46.4	6.10	21.0	0.41	2.37	0.67	1.63
1.0	14.5	27.9	42.8	—	—	—	2.07	—	—
1.5	6.4	27.6	48.2	7.14	17.1	0.46	2.36	0.52	1.13
2.0	6.5	28.0	31.9	7.94	32.1	0.50	1.54	0.96	1.91
3.0	19.6	28.7	34.8	8.00	28.5	0.49	1.64	0.83	1.69
3.0	6.2	26.3	39.9	—	—	—	2.05	—	—
4.0	6.6	28.0	35.2	11.02	25.7	0.70	1.70	0.77	1.10

Copper and Rose /14/ studied the reaction between alkylaluminum compounds and titanium tetrachloride and found a relationship between the amount of the liberated alkane and the degree of reduction of the titanium.

This reaction may also be followed by the method of electron paramagnetic resonance (EPR). It was found /24/ that the system gives a sharp EPR signal some time after the reactants have been intermixed. It was also noted that an EPR signal is not given by the precipitate, which consists mainly of $TiCl_3$ formed by the reduction of $TiCl_4$. According to the authors, this may be due to the rapid spin-lattice relaxation of Ti^{3+} ions. In another study, these workers investigated the nature of the EPR signal given by the precipitate formed by the reaction between diethylaluminum chloride and titanium tetrachloride in heptane /25/. It was found that a certain time after the reactants have been mixed together, an EPR (singlet) signal appears. This signal is not very intense and consists of two superposed singlets:

symmetrical, with g-factor 1.925,* and a weaker asymmetrical one, with g-factor 1.933. The signal intensity increases rapidly at first, and then much less rapidly, as the precipitate separates out. It was also noted that the precipitate itself gave an EPR signal. The authors concluded that the EPR signals were given by Ti^{3+} ions, which originate from complex formation between $Al(C_2H_5)_2Cl$ and $TiCl_4$, which is followed by the reduction of the titanium in the complex to Ti^{3+}. It was found that the higher the Lewis acid strength of the alkylaluminum compound, the higher the concentration of this soluble complex and the higher the intensity of the EPR signal.

Varodi et al. /26/ attempted to clarify the nature of the catalytic effect of the complex and of the interaction between the constituents. He confirmed that the initial reaction which takes place in the system $Al(C_2H_5)_2Cl - TiCl_4$ is the alkylation of titanium tetrachloride to $C_2H_5TiCl_3$. The ethyltitanium trichloride, which precipitates out 30 minutes after the intermixing of the components, contains no aluminum and gives no EPR signal at room temperature. However, the precipitate of $Ti_3Cl_6Al(C_2H_5)$, which was collected 4 hours after the intermixing, contained trivalent titanium and gave an EPR signal. These workers also found that the precipitates were sparingly soluble in hydrocarbons, but were about 1000 times more soluble in the presence of titanium tetrachloride. They found that the catalytic effect does not originate from any particular compound obtained by reacting $Al(C_2H_5)_2Cl$ with $TiCl_4$, but from the system: precipitate — soluble complex, as a whole. The presence of $TiCl_4$ in the system is particularly important, assisting as it does in the formation of the maximum amount of the soluble compound, which is the active co-catalyst.

Attempts have been made to follow the reaction of alkylaluminum compounds with titanium tetrachloride by means of IR spectroscopy. One of the first studies on the subject is that of Groenewege /27/, who studied the IR spectra of the system $(CH_3)_2AlCl - TiCl_4$ in heptane. Gray et al. /28, 29/ subsequently succeeded in taking the IR spectra of the initial and of the intermediate compounds and thus to follow the course of the reaction between trimethylaluminum and titanium tetrachloride, despite the very high reactivities of both the reactants and the reaction products. The study of the IR spectra of these compounds and their reaction products at $150 - 300\ cm^{-1}$ and also on both sides of this range failed to reveal absorption bands corresponding to these products. The reactants were introduced into the cells in the gaseous state. Figure 4 shows the initial spectra of the reactant mixtures at different molar ratios $Al(C_2H_5)_3 : TiCl_4$. Inspection of these spectra in conjunction with the IR spectra of individual compounds led the authors to the conclusion that when the reactants are present in equimolar ratio, the initial reaction is

$$(CH_3)_3Al + TiCl_4 \longrightarrow (CH_3)_2AlCl + CH_3TiCl_3$$

If this ratio is lower, $TiCl_4$ is also to some extent alkylated by the dimethylaluminum chloride; it is probable that the result is the establishment of the following equilibrium:

$$(CH_3)_2AlCl + TiCl_4 \rightleftarrows CH_3AlCl_2 + CH_3TiCl_3$$

* The g-factor is a parameter which describes the spin-orbital bond. For the free electron g = 2.0033.

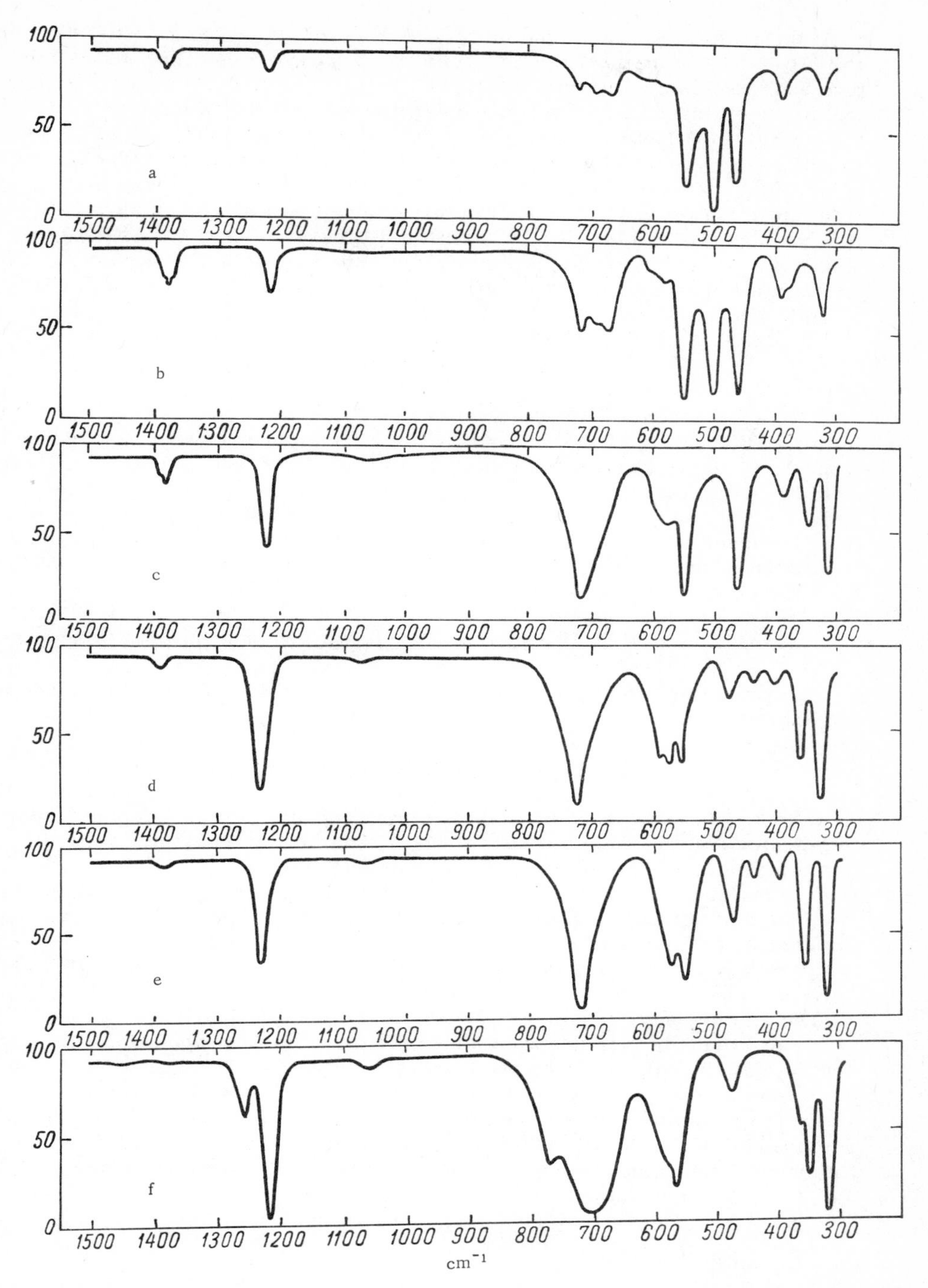

FIGURE 4. Infrared spectra of reactant mixtures formed during the reactions between $TiCl_4$ and trimethylaluminum: a — molar ratio Al : Ti = 1 : 3; b — 2 : 3; c — 1 : 1; d — 2 : 1; e — 8 : 3; f — 4 : 1.

If the ratio is smaller than unity, $TiCl_4$ is alkylated by trimethylaluminum to $(CH_3)_2TiCl_2$; if trimethylaluminum is present in excess, the alkylation proceeds to trimethyltitanium chloride, but this compound is unstable and has not been identified directly. Unreacted trimethylaluminum remains behind only if it has been introduced in a large excess (4:1). It was found that if the reaction is carried out in the gas phase, titanium tetrachloride is not fully alkylated. If the ratio $Al(CH_3)_3:TiCl_4$ is 1:1 or less, methyltitanium trichloride is photochemically decomposed at a very slow rate (the decomposition lasts for 50 to several hundred hours). If $Al(CH_3)_3:TiCl_4 = 1:3$, all the initial trimethylaluminum is eventually consumed with formation of CH_3TiCl_3 and CH_3AlCl_2, which reacts with the residual $TiCl_4$ as follows:

$$CH_3AlCl_2 + TiCl_4 \rightleftarrows CH_3TiCl_3 + AlCl_3$$

This equilibrium is strongly shifted to the left; this was shown by reacting pure CH_3TiCl_3 with $AlCl_3$. These results /29/ remain valid whether the reaction is realized in the gas phase or in hydrocarbon solutions, and also whether trimethylaluminum or another alkylaluminum compound, with a longer hydrocarbon chain, is employed. It was also noted that the thermochemical and photochemical reactions of methyltitanium trichloride result in the formation of the brown-colored modification of titanium trichloride and hydrocarbons:

$$CH_3TiCl_3 \longrightarrow CH_3\cdot + TiCl_3$$

Disproportionation accompanied by the formation of $TiCl_4$ takes place at the same time:

$$2CH_3TiCl_3 \longrightarrow (CH_3)_2TiCl_2 + TiCl_4$$

It was shown by Beermann and Bestian /30/ that at high temperatures the mechanism of decomposition of methyltitanium trichloride is more complex and the reaction is probably autocatalytic.

A study of the reaction between trimethylaluminum and dimethylaluminum chloride and titanium tetrachloride by the method of NMR spectra was effected by Sakurada et al. /32/. This reaction was studied at room temperature, with equimolar reactant ratio, in tetrahydrofuran (THF) as solvent.

The authors also concluded /32/ that even if methyltitanium trichloride is in fact formed, it is unstable under the experimental conditions employed and rapidly decomposes with the formation of $TiCl_3 \cdot 3THF$ and CH_4. They discarded a possible formation of compounds of bivalent titanium.

On the strength of these results and also of the IR spectra of individual compounds dissolved in tetrahydrofuran, these authors proposed the following mechanism of the reaction between methylaluminum compounds and $TiCl_4$ in tetrahydrofuran:

$$Al(CH_3)_3 + TiCl_4 \xrightarrow{THF} Al(CH_3)_2Cl + TiCl_3 + CH_3\cdot$$
$$Al(CH_3)_2Cl + TiCl_4 \xrightarrow{THF} Al(CH_3)Cl_2 + TiCl_3 + CH_3\cdot$$
$$\longrightarrow CH_4$$

This reaction scheme was confirmed by the results of Bohar et al. /32a/.

2. REACTIONS OF ALKYLALUMINUM COMPOUNDS WITH TITANIUM TRICHLORIDE

Reactions between alkylaluminum compounds and titanium trichloride are much slower than their reactions with titanium tetrachloride. The nature of the reactions between these compounds, which result in the formation of the catalyst complex, was until recently unclear. According to some of the workers, the main reaction consisted in the sorption of the trialkylaluminum compound on the surface of titanium trichloride during the formation of the catalyst complex. According to others, chemical reactions took place at elevated temperatures only /15, 31, 33, 37/. Thus, it was shown by Boldyreva et al. /15/, who studied the reaction between triethylaluminum and the violet modification of titanium trichloride, that this modification reacts with triethylaluminum only very sluggishly. The brown modification of $TiCl_3$, which is formed by the reaction between titanium tetrachloride and alkylaluminum compounds, is much more reactive with respect to triethylaluminum than the violet modification, due to the differences in the crystal structure of the two modifications. The reaction between triisobutylaluminum and titanium trichloride in heptane at varying reactant ratios yielded solid products in which the contents of titanium and chlorine were approximately equal.

However, the content of aluminum and of alkyl groups increased with increasing $Al(iso\text{-}C_4H_9)_3:TiCl_3$ ratio /31/. Very probably, the molecules of the alkylaluminum compounds are first adsorbed on the crystals of $TiCl_3$, after which the decomposition of the complex $TiCl_3 \cdot AlR_3$ takes place very slowly, as follows:

$$TiCl_3 \cdot AlR_3 \longrightarrow TiCl_2 + R \cdot AlR_2Cl$$

Similar results were obtained by Natta, who worked with triethylaluminum tagged with ^{14}C in the ethyl group /38/.

Simon et al. /39/ studied this reaction in more detail (he also made a detailed study of the reaction between alkylaluminum compounds and titanium tetrachloride). They concluded that $TiCl_3$ reacts with triethylaluminum both at elevated temperatures (70 — 110°C) and, to a certain extent, at room temperature. The product formed as a result of this reaction was found to catalyze the polymerization of olefins. The authors found that the reaction temperature, the time during which $TiCl_3$ is being ground and the molar ratio of the reactants affect the composition of the precipitate only during the first 4 — 10 hours. Thereafter, the composition of the precipitate remains practically unchanged. They also showed that the reaction between triethylaluminum and titanium trichloride is accompanied by gas evolution.

A study of the degree of reduction of titanium as a result of the reaction between triethylaluminum and $TiCl_3$ showed that titanium was reduced to the metal, especially so if the molar ratio of the reactants was high, if the $TiCl_3$ was finely dispersed and if the reaction temperature was high. According to the authors, this is caused by the ease with which bivalent titanium passes into the metallic state. Very probably, Ti^{2+} is reduced to Ti^{+}, which then disproportionates:

$$2Ti^{+} \longrightarrow Ti^{2+} + Ti^{0}$$

This is why monovalent titanium in the precipitate cannot be analytically determined.

Data obtained on the extent of the reduction of $TiCl_3$ show that this depends, first and foremost, on the reactant ratio and increases both with the reaction temperature and with the degree of dispersity of titanium trichloride. That titanium is reduced is also indicated by the gas evolution which takes place during the reaction. The reaction scheme is similar to that given above for triethylaluminum and titanium tetrachloride.

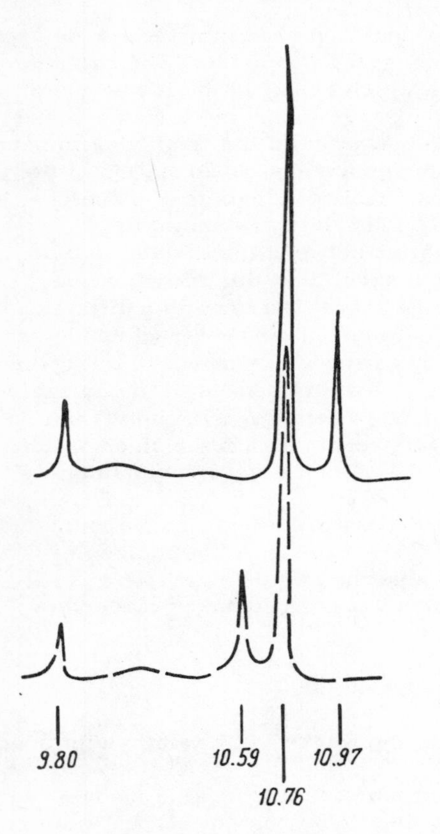

FIGURE 5. NMR spectra of a mixture of $Al(CH_3)_3$ (continuous line) or $Al(CH_3)_2Cl$ (dotted line) and titanium trichloride in tetrahydrofuran (concentration of AlR_2X, 0.09 M; equimolar amounts of AlR_2X and $TiCl_3$).

According to the authors of the above paper, the main difference between the two reactions consists in the fact that the reaction with the trichloride takes place only on the surface of the crystal lattice of $TiCl_3$, and that the trichloride is able to react with the alkylaluminum compound only gradually, as the crystal structure becomes disrupted.

Simon et al. /39/ postulated, on the strength of their experimental results, that aluminum is directly bound to titanium in the complex of alkylaluminum compound with $TiCl_3$. They base this assumption on the fact that the analytically determined content of ethyl groups is, very nearly, two ethyl groups per aluminum atom, i. e., the catalyst contains an $-Al(C_2H_5)_2$ group. Titanium is also directly bound to chlorine atoms. However, this conclusion is, at best, only a working hypothesis in the study of the composition, structure and activity of the catalyst complex. Similar results were also reported by other workers /39a/.

The reaction between trialkylaluminum compounds and titanium trichloride was also studied by methods other than chemical analysis. Sakurada et al. /32/ carried out one of the first studies in which the reaction products were identified by means of NMR spectra. These workers gave interpretations of NMR spectra of a number of methyl derivatives of aluminum and their products of reaction with $TiCl_4$, and also took and interpreted NMR spectra of the reaction between trimethylaluminum and titanium trichloride. The use of tetrahydrofuran as solvent distorted the true picture somewhat, owing to the formation of the soluble complexes $Al(CH_3)_3 \cdot THF$ and $TiCl_3 \cdot 3THF$, but the main features of the reaction could nevertheless be studied.

The reaction between trimethylaluminum and $TiCl_3$ was studied at room temperature and molar ratios $Al(CH_3)_3:TiCl_3$ of 1:1, 2:1 and 3:1. The NMR spectra of these systems were taken 20 hours after the components had been mixed. To do this, a saturated (0.1 M) concentration of $TiCl_3 \cdot 3THF$ in THF was prepared. No precipitate separated out as a result of the reaction, but the solution changed color from violet to blue. When $Al(CH_3)_2Cl$ was employed, the color changed to brown. Figure 5 shows the NMR spectra of solutions obtained by reacting $Al(CH_3)_3$ with $TiCl_3$ in equimolar ratio. These spectra greatly resemble those obtained for the reaction between $Al(CH_3)_3$ and $TiCl_4$. In both cases a weak signal is noted at $\tau = 9.80$. Sakurada et al. attributed this signal to the proton of methane. They pointed out that a similar signal of an equal intensity was noted in an 0.09 M solution of $Al(CH_3)_3$, which suggests the formation of methane as a result of the reaction between $Al(CH_3)_3$ and the water in the solvent. The location of NMR signals remains more or less constant when the molar ratio between the reactants is varied (Al:Ti = 2:1 and 3:1). However, the intensities of the signals attributable to the protons of methyl radicals in $Al(CH_3)_2Cl$ depends on the molar ratio of the initial components. An increase in the molar ratio of $Al(CH_3)_3$ to $TiCl_3$ increases the surface area of the signal at $\tau = 10.97$ to that of the signal at $\tau = 10.76$. It was concluded by the authors /32/ that if the reactant ratio Al:Ti is higher than 1:1, not all the trimethylaluminum reacts. Basing themselves on experimental determinations of magnetic susceptibility the authors maintain that significant amounts of bivalent titanium are not formed in these reaction mixtures.

They concluded that the mechanism of the reactions of $Al(CH_3)_3$ and $Al(CH_3)_2Cl$ with titanium trichloride can be described by the following equations:

$$Al(CH_3)_3 + TiCl_3 \xrightarrow{THF} Al(CH_3)_2Cl + TiCl_2CH_3$$

$$Al(CH_3)_2Cl + TiCl_3 \xrightarrow{THF} Al(CH_3)Cl_2 + TiCl_2CH_3$$

No NMR signal of the proton in $TiCl_2CH_3$ was obtained, so that the question of the existence of this compound under these conditions must remain open.

3. REACTIONS BETWEEN ALKYLALUMINUM COMPOUNDS AND BIS-CYCLOPENTADIENYLTITANIUM DICHLORIDE

The first attempts to study the reactions between organoaluminum compounds and $(C_5H_5)_2TiCl_2$ are due to Natta et al. /40/, who suggested that the resulting products be utilized in the study of the polymerization of α-olefins. When 0.01 mole of $(C_5H_5)_2TiCl_2$, suspended in 50 ml heptane, was mixed with 0.025 mole of $Al(C_2H_5)_3$ and the mixture was allowed to stand for one hour, a blue-colored solution was obtained. The reaction temperature did not exceed 70°C. A blue-colored crystalline compound was obtained when the solution was cooled to −50°C. After several recrystallizations from heptane the product had a m. p. of 126 — 130°C and a composition corresponding to the formula $(C_5H_5)_2TiCl_2Al(C_2H_5)_2$. The complex was subsequently studied by X-ray spectroscopy /41/. It was found that the atoms of titanium and aluminum are coplanar and are connected to each other by means of a four-membered ring formed by chlorine atoms. The distance between the Ti and Cl atoms, and also between the Al and Cl atoms is about 2.5 Å. The plane of the C_5H_5-group is perpendicular to the line connecting the titanium atom with the center of the group; all five Ti — C bonds are about 2.3 Å long. The ethyl groups are bound to aluminum atoms. It was also noted that the cyclopentadienyl groups in the complex are not parallel to one another, as in ferrocene, but are somewhat displaced with respect to one another, owing to the type of hybridization of the metal atom to which they are bound. Figure 6 is a schematic representation of the structural model of the complex $[(C_5H_5)_2TiAl(C_2H_5)_2]_2$ /41a/.

It was established by Breslow and Newburg /42, 43/ that this formula is more correctly written as $(C_5H_5)_2TiCl \cdot Al(C_2H_5)_2Cl$. These workers subsequently studied the reactions of $(C_5H_5)_2TiCl_2$ with diethylaluminum chloride and triethylaluminum. The reactions were performed both in toluene, in which $(C_5H_5)_2TiCl_2$ is readily soluble, and in heptane. They found that when a toluene solution of $(C_5H_5)_2TiCl_2$ is mixed with $Al(C_2H_5)_2Cl$, the color of the reaction mixture rapidly changes from orange to dark red and then gradually turns green and finally blue. According to the analytical data, the blue-colored complex in toluene corresponds to the compound $(C_5H_5)_2TiCl \cdot {}^1/_2Al_2(C_2H_5)_3Cl_3$, m.p. 80 — 90°C. The authors /42, 43/ did not consider the above composition as conclusively established, but were convinced that the titanium in the compound was trivalent. This was shown by the fact that the hydrolysis of the reaction product by aqueous sulfuric acid yielded a green-colored solution, in which the titanium could be titrated as the trivalent ion. When the blue complex was heated with ether, green crystals separated out; their composition was established by analysis as $(C_5H_5)_2TiCl$. Determinations of magnetic susceptibility showed the presence of one unpaired electron.

In one of his subsequent studies /44/ Breslow discussed the spectroscopic evidence in favor of the existence of three complexes formed during the reaction between $(C_5H_5)_2TiCl_2$ and $Al(C_2H_5)_2Cl$. Figures 7 and 8 represent the changes in the visible spectrum which occur when toluene solutions of $(C_5H_5)_2TiCl_2$ and $Al(C_2H_5)_2Cl$ are mixed together in the molar ratio 1:1.5 at 2.3°C. It was found that the spectrum gradually varied with the duration of the interaction of the components. It was reported by Breslow that the formation of the complex $(C_5H_5)_2TiCl_2 \cdot Al(C_2H_5)_2Cl$ was rapid and that it was then gradually converted to the complex $(C_5H_5)_2Ti(C_2H_5)Cl \cdot Al(C_2H_5)Cl_2$ (spectrum III), which is relatively stable at low temperatures. According to these workers, this is the complex which is responsible for the polymerization of olefins. The formation of $(C_2H_5)_2TiCl \cdot Al(C_2H_5)Cl_2$ is slow (Figure 8), and this complex is not active during the polymerization of olefins. Basing himself on these data, Breslow considers that the reaction between bis-cyclopentadienyltitanium dichloride and diethylaluminum chloride may be described in the following manner.

1. Formation of complex:

$$(C_5H_5)_2TiCl_2 + Al(C_2H_5)_2Cl \longrightarrow (C_5H_5)_2Ti{<}^{Cl}_{Cl}{>}Al(Cl)(C_2H_5)_2$$

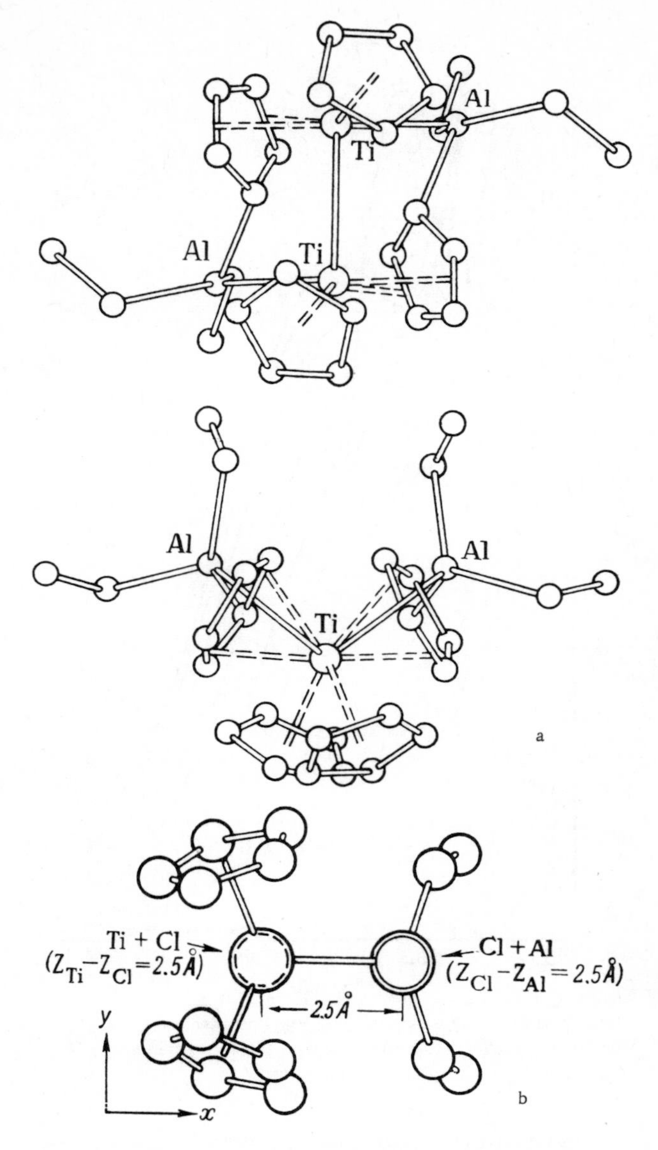

FIGURE 6. Structure of complexes $[(C_5H_5)_2TiAl(C_2H_5)_2]_2$, (a) and π-$(C_5H_5)_2TiCl_2Al(C_2H_5)_2$, (b).

2. Alkylation of the titanium compound:

$$(C_5H_5)_2Ti{\begin{matrix} Cl \\ Cl\cdots Al(C_2H_5)_2Cl \end{matrix}} \longrightarrow (C_5H_5)_2Ti{\begin{matrix} C_2H_5 \\ Cl\cdots Al(C_2H_5)Cl_2 \end{matrix}}$$

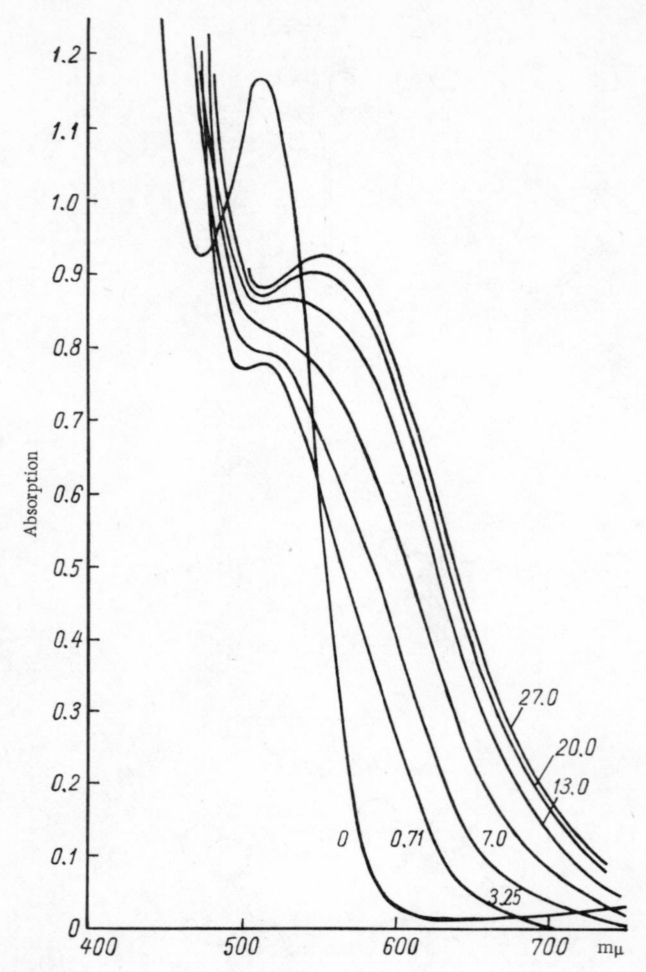

FIGURE 7. Evolution of the visible spectrum of the reaction mixture, consisting of 5 mmoles of $(C_5H_5)_2TiCl_2$ and 7.5 mmoles of $(C_2H_5)_2AlCl$ at 2.3° C (numbers on curves indicate times in minutes after the mixing of the components).

3. Formation of trivalent titanium compound:

$$(C_5H_5)_2Ti(C_2H_5)Cl \cdot Al(C_2H_5)Cl_2 \longrightarrow (C_5H_5)_2TiCl \cdot Al(C_2H_5)Cl_2 + C_2H_5 \cdot$$

Kissin et al. /45/ used IR spectroscopy to confirm the formation of blue complexes of the type:

$$(C_5H_5)_2Ti\begin{matrix} Cl \\ Cl \end{matrix}Al(C_2H_5)_2 \text{ and } (C_5H_5)_2Ti\begin{matrix} Cl \\ Cl \end{matrix}Al(C_2H_5)Cl$$

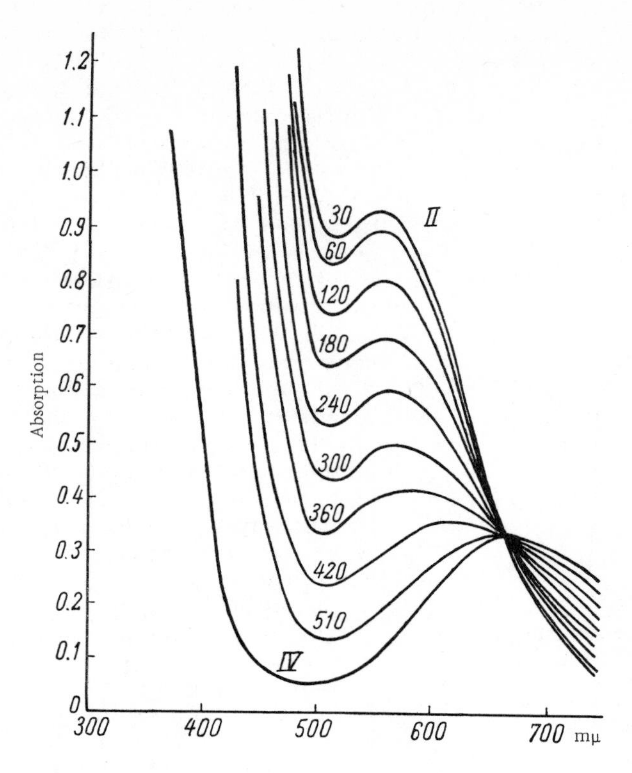

FIGURE 8. Evolution of the visible spectrum of the reaction mixture consisting of 5 mmoles of $(C_5H_5)_2TiCl_2$ and 7.5 mmoles of $(C_2H_5)_2AlCl$ (numbers on curves indicate times in minutes after the mixing of the components).

Fushman et al. /55/ made an interesting study of the effect of different solvents on the interaction of $(C_5H_5)_2TiCl_2$ with $Al(C_2H_5)_2Cl$. Whereas titanium in $(C_5H_5)_2TiCl_2$ is readily reduced to Ti(III) by triethylaluminum dissolved in heptane and the reduction is arrested at this stage, the amount of the reduced titanium first increases and then gradually decreases if dichloroethane is employed as solvent. The authors maintain that redox processes take place in the complex being formed, as a result of which the amount of the organoaluminum component decreases to zero. According to them, the oxidation of Ti(III) derivatives is effected by dichloroethane. When benzene or chlorobenzene was used as solvent, the concentration of Ti(III) in the complex formed did not decrease.

Organometallic compounds of the general formula $R_2Ti[CH_2Al(CH_3)X]_2$, where R is cyclopentadienyl or alkylcyclopentadienyl and X is a halogen or an alkoxy group, were recently prepared. These substances act as igniters of pyrotechnic articles, as sources of free radicals and polymerization catalysts /35/.

The structure of the products of the reaction between alkylaluminum compounds and bis-cyclopentadienyltitanium dihalides was investigated by Shilov et al. /46, 47/ by means of EPR spectra. When trimethylaluminum, triethylaluminum, triisopropylaluminum, triisobutylaluminum, triphenylaluminum or diethylaluminum chloride were mixed with $(C_5H_5)_2TiCl_2$, $(C_5H_5)_2TiBr_2$ or $(C_5H_5)_2TiI_2$ in equimolar ratios of the initial components, similar EPR signals were obtained with a g-factor of 1.975 (Figure 9), which was attributed by the authors to the presence of unresolved hyperfine structure. The study of EPR spectra obtained at other $AlR_3:(C_5H_5)_2TiCl_2$ ratios revealed the individual features of the spectra. It was found that at Al:Ti ratios higher than unity triethylaluminum, triisopropylaluminum and triisobutylaluminum react with $(C_5H_5)_2TiCl_2$ to give new EPR signals with a highly resolved hyperfine structure, while the spectra of complexes of trimethylaluminum, triphenylaluminum or diethylaluminum chloride with the titanium compound remain unchanged. When the Al:Ti ratio in the former group of complexes was varied from 1:1 to about 20:1, signal I (Figure 9) changed to signal II (Figure 10), which is a doublet with a g-factor of 1.985. When the Al:Ti ratio was further increased to 50:1, signal II changed to signal III with a g-factor of 1.988, consisting of 8 components (Figure 11). In these experiments the shapes of signals II and III were independent of the identity of the alkyl group in the organoaluminum compound and of the identity of the halogen atoms in the cyclopentadienyltitanium compound. Changes of signals from I to III were also noted at various concentrations of the reaction components.

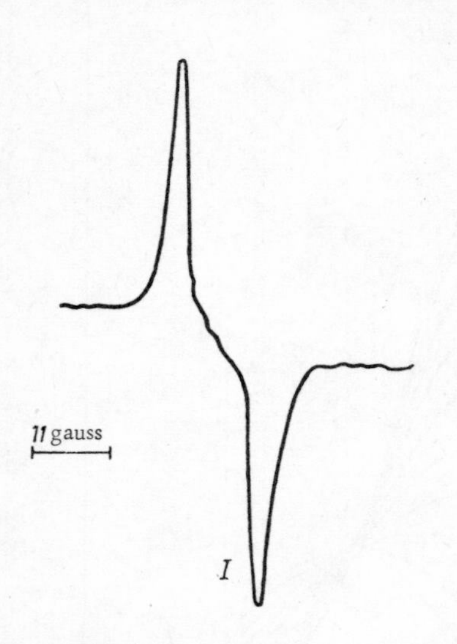

FIGURE 9. EPR spectrum of the mixture of alkylaluminum compounds with $Ti(C_5H_5)_2X_2$ in toluene (equimolar ratio of Al to Ti; reagent concentration less than $1 \cdot 10^{-3}$ mole/liter).

The authors concluded that their alkylaluminum compounds contained dialkylaluminum hydride; the existence of this hydride was subsequently experimentally confirmed. This is the reason for the changes in the EPR spectra of these complexes. According to the authors, the hydrogen of the dialkylaluminum hydride acts as the bridge connecting the alkylaluminum compound with $(C_5H_5)_2TiCl_2$; this is the cause for the formation of the hyperfine structure in spectrum III. The participation of the hydride hydrogen in the complexes thus formed was also confirmed by experiments involving the deuterated aluminum compound $AlD[CH_2C(CH_3)_2]_2$. The EPR spectra of the reaction products of this compound with $(C_5H_5)_2TiCl_2$ corresponded to signals II and III.

These and other experimental data led Shilov et al. /46, 47/ to propose the following scheme for the reactions between $(C_5H_5)_2TiCl_2$ and dialkylaluminum hydrides:

$$AlR_2H + (C_5H_5)_2TiCl_2 \longrightarrow (C_5H_5)_2Ti\langle{}^{Cl}_{Cl}\rangle AlR_2 + H\cdot$$

$$(C_5H_5)_2Ti\langle{}^{Cl}_{Cl}\rangle AlR_2 \rightleftarrows (C_5H_5)_2TiCl + AlR_2Cl$$

$$(C_5H_5)_2TiCl + AlR_2H \rightleftarrows (C_5H_5)_2Ti\langle {}^{H}_{Cl} \rangle AlR_2$$

The compound $(C_5H_5)_2Ti\langle {}^{H}_{Cl} \rangle AlR_2$ contains a hydrogen bridge in lieu of one chlorine bridge. This bridge is responsible for the doublet in signal II. Further substitution in this manner yields a compound with two hydrogen bridges $(C_5H_5)_2Ti\langle {}^{H}_{H} \rangle AlR_2$, which are responsible for signal III.

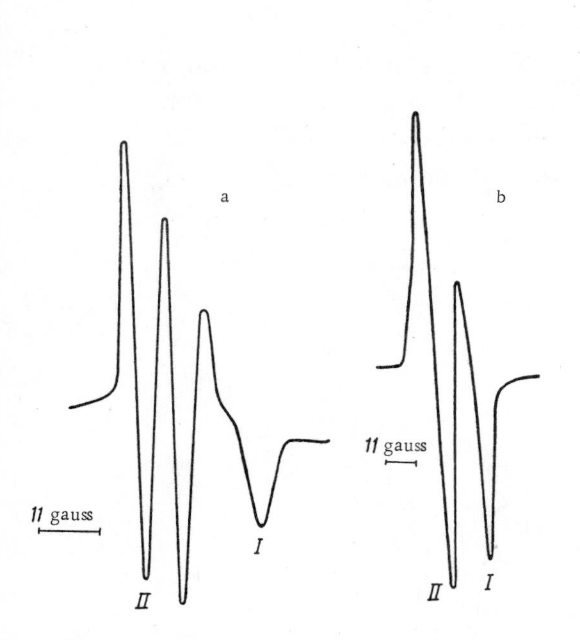

FIGURE 10. EPR spectra of the mixture of AlR_3 with $Ti(C_5H_5)_2X_2$, containing complex compounds of the structure $(C_5H_5)_2Ti\langle {}^{H}_{X} \rangle AlR_2$:

a — molar ratio $AlR_3 : Ti(C_5H_5)_2X_2 = 20:1$; b — molar ratio $AlD[CH_2CD(CH_3)_2] : Ti(C_5H_5)_2Cl_2 = 2:1$.

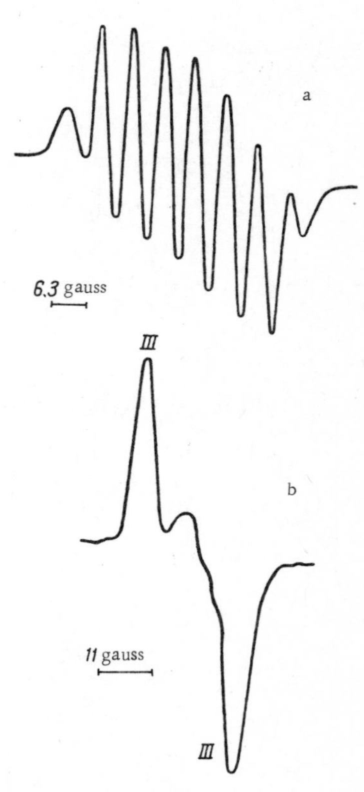

FIGURE 11. EPR spectra of the mixture of AlR_2H with $Ti(C_5H_5)_2Cl_2$ at a molar ratio $Al:Ti = 3:1$, containing a complex compound of the structure $(C_5H_5)_2Ti\langle {}^{H}_{H} \rangle AlR_2$.

The shift of the g-factor from signal I to signal III shows that the nature of the orbital of the unpaired electron changes. According to the authors /46, 47/ the spin density of the unpaired electron on the aluminum atom increases during this change. Subsequently, the mechanism of interaction of these reagents, based on the kinetic relationships governing the reactions between organoaluminum compounds and titanium halides was proposed by Zefirova and Shilov /48/. They studied the products of the reactions between trimethylaluminum and $(C_5H_5)_2TiCl_2$, between triethylaluminum and $(C_5H_5)_2TiCl_2$ and $TiCl_4$ and between $Al(C_2H_5)_2Cl$ and $(C_5H_5)_2TiCl_2$. It was found that in the reaction between trimethylaluminum and $(C_5H_5)_2TiCl_2$ the reduction of titanium proceeds slowly and the complex $(C_5H_5)_2TiCl_2 \cdot Al(CH_3)_3$ is not formed. Only in the presence of an olefin (ethylene, propylene and butylene were studied) did the red solution turn blue, with evolution of methane and absorption of olefin. The blue complex thus formed was distilled; its structure was established as $(C_5H_5)_2TiCl \cdot Al(CH_3)_2Cl$ on the strength of its visible and EPR spectra.

Basing themselves on certain kinetic relationships, the authors proposed the following mechanism for the reaction between diethylaluminum chloride and bis-cyclopentadienyltitanium dichloride:

$$[Al(C_2H_5)_2Cl]_2 \rightleftarrows Al(C_2H_5)_2^+ + Al(C_2H_5)_2Cl_2^-$$

$$2(C_5H_5)_2TiCl_2 + [Al(C_2H_5)_2Cl]_2 \rightleftarrows 2(C_5H_5)_2TiCl_2 \cdot Al(C_2H_5)_2Cl$$

$$(C_5H_5)_2TiCl_2 \cdot Al(C_2H_5)_2Cl \rightleftarrows (C_5H_5)_2Ti(C_2H_5)Cl \quad Al(C_2H_5)Cl_2$$

$$(C_5H_5)_2Ti(C_2H_5)Cl \cdot Al(C_2H_5)Cl_2 \longrightarrow [(C_5H_5)_2TiC_2H_5]^+ + Al(C_2H_5)Cl_3^-$$

$$[(C_5H_5)_2TiC_2H_5]^+ + Al(C_2H_5)_2Cl_2^- \longrightarrow (C_5H_5)_2Ti(C_2H_5)_2 + Al(C_2H_5)Cl_2$$

$$(C_5H_5)_2Ti(C_2H_5)_2 + Al(C_2H_5)_2^+ \longrightarrow [(C_5H_5)_2Ti(C_2H_5)]^+ + Al(C_2H_5)_3$$

$$(C_5H_5)_2Ti(C_2H_5)_2 \longrightarrow (C_5H_5)_2Ti + C_2H_6 + C_2H_4$$

$$(C_5H_5)_2Ti + (C_5H_5)_2TiCl_2 \cdot Al(C_2H_5)_2Cl \longrightarrow (C_5H_5)_2TiCl + (C_5H_5)_2TiCl_2Al(C_2H_5)_2$$

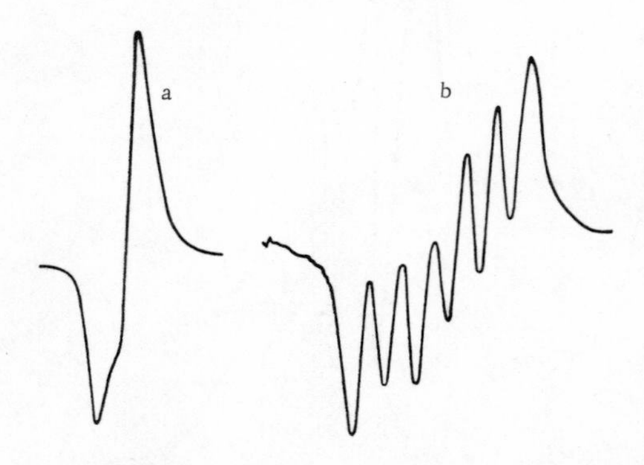

FIGURE 12. EPR spectra of the products of the reaction of $(C_5H_5)_2TiCl_2$ with $Al(CH_3)_2Cl$ (a) and with $Al(CH_3)Cl_2$ (b) in the presence of heptane.

We have already said that the presence of an olefin considerably accelerates the formation of the blue complex of $(C_5H_5)_2TiCl_2$ with $Al(CH_3)_3$. This effect was studied by Shilov et al. /49/ who found that the addition of an α-olefin results in a rapid conversion of the complex $(C_2H_5)_2Ti(CH_3)Cl \cdot Al(CH_3)Cl_2$, which is stable for several hours. The product is blue, contains trivalent titanium and has been identified by spectroscopic analysis as the complex $(C_5H_5)_2TiCl \cdot AlR_2Cl$. Figure 12 shows the EPR spectrum of this complex, with a g-factor of 1.975, and signal intensity of 20.4 oersteds. These parameters are close to those obtained for the product of the reduction of $(C_5H_5)_2TiCl_2$ by diethylaluminum chloride /47, 50/. If this reduction takes place in the presence of an olefin, titanium is also reduced to Ti(III), but to a much smaller extent. The g-factor of this EPR spectrum is 1.975, and the spectrum has a well-resolved hyperfine structure consisting of six lines of equal intensities (Figure 12b); this indicates that there is an interaction between the lone electron in the complex $(C_5H_5)_2TiCl \cdot AlR_2Cl$ and the aluminum nucleus (spin 5/2); the width of the signal is 37.5 oersteds. These workers also showed that the electrical conductivity of the benzene solution of $Al(CH_3)_2Cl$ increases several times on the addition of $(C_5H_5)_2TiCl_2$, whereas neither of these solutions is markedly conducting by itself. It was established that the electrical conductivity of the complex is proportional to the square root of its concentration. The authors used these data to confirm the ionic nature of the reaction between the constituents of the complex (p. 102) and found that, in the presence of an olefin, the rate-determining stages are the reactions between the olefin and $(C_5H_5)_2Ti^+CH_3$ or between $(C_5H_5)_2Ti^+R$ and the ion $Al(CH_3)_2Cl_2^-$:

$$(C_5H_5)_2Ti(CH_3)Cl \cdot Al(CH_3)Cl_2 \rightleftarrows (C_5H_5)_2Ti^+CH_3 + Al(CH_3)Cl_3^-$$

$$(C_5H_5)_2Ti^+CH_3 + C_7H_{14} \longrightarrow (C_5H_5)_2Ti{-}CH_2{-}\underset{\underset{C_5H_{11}}{|}}{CH}{-}CH_3$$

$$(C_5H_5)_2Ti^+CH_2{-}CH(CH_3)C_5H_{11} + Al(CH_3)_2Cl_2^- \longrightarrow Al(CH_3)Cl_2 + \\ + (C_5H_5)_2Ti(CH_3)CH_2{-}CH(CH_3)C_5H_{11}$$

$$(C_5H_5)_2Ti(CH_3)CH_2{-}CH(CH_3)C_5H_{11} \longrightarrow (C_5H_5)_2Ti + CH_4 + \\ + CH_2{=}C(CH_3)C_5H_{11}$$

$$(C_5H_5)_2Ti + (C_5H_5)_2TiCl_2 \cdot Al(CH_3)_2Cl \longrightarrow (C_5H_5)_2TiCl + \\ + (C_5H_5)_2TiCl \cdot Al(CH_3)_2Cl \text{ etc.}$$

Thus, the olefin molecule attacks the $Ti-CH_3$ bond of the complex and then, immediately following this reaction, the alkyl group formed disproportionates with the methyl group; the titanium is reduced and loses its activity.

Shilov et al. subsequently produced an experimental confirmation of his mechanism of the reaction between $(C_5H_5)_2TiCl_2$ and $Al(CH_3)_2Cl$ in the presence of ethylene. He studied the changes in the spectra of the catalyst systems in the presence of small amounts of ethylene /51/ and showed that the propagation of the polymer chain involves attack on the $Ti-C$ bond, since the methyl groups of $Ti-CH_3$ in the complex are gradually converted to propyl, pentyl, etc., groups as follows:

$$Ti{-}CH_3 \xrightarrow{C_2H_4} Ti{-}CH_2{-}CH_3 \xrightarrow{C_2H_4} Ti{-}(CH_2{-}CH_2)_2{-}CH_3 \text{ etc.}$$

This mechanism was confirmed by Grigoryan and D'yachkovskii et al. who studied the identity of the active sites of the homogeneous catalytic system constituted by $(C_5H_5)_2TiCl_2$ with $Al(CH_3)_2Cl$ /51a/.

Basing themselves on Bartelink's /52/ method for the determination of the $Ti-CH_3$ group by nuclear magnetic resonance, D'yachkovskii et al. /53/ showed that the concentration of $Ti-CH_3$ groups in the catalyst complex gradually decreased, while the concentration of $Ti-CH_2-CH_2-CH_3$ groups increased. The latter groups could only have been formed following the attack of the $Ti-CH_3$ bond, formed by the reaction between $(C_5H_5)_2TiCl_2$ and $Al(CH_3)_2Cl$, with ethylene. Similar results were also obtained when ethylene was substituted by deuteroethylene /54/. These workers recently published /54a/ a survey of the reactions involved and of the operative mechanism of soluble complex catalysts.

4. REACTIONS OF ALKYLALUMINUM COMPOUNDS WITH COMPOUNDS OF VANADIUM AND OTHER TRANSITION METALS

The composition and certain catalytic properties of the products of the reaction between vanadium chlorides and alkylaluminum compounds were first studied by Natta et al. /56 — 61/. They pointed out, in particular, that when vanadium oxychloride is reduced by an alkylaluminum compound, the resulting precipitate mainly consists of compounds of vanadium chlorides with organoaluminum compounds. They also showed that the content of chlorine in the precipitate decreases with the increase in the proportion of the organoaluminum compound in the initial reactant mixture. Petrov and Korotkov /62/ studied the reaction between vanadium oxytrichloride and triethylaluminum; the relationships experimentally found made it possible to predict the composition of the reaction products by calculation. They showed that the reaction between these two compounds initially proceeds in two ways:

1. Alkylation by attack on $V-Cl$ bond:

$$VOCl_3 + Al(C_2H_5)_3 \longrightarrow C_2H_5VOCl_2 + Al(C_2H_5)_2Cl$$

$$C_2H_5VOCl_2 \longrightarrow VOCl_2 + C_2H_5\cdot$$

2. Alkylation by attack on $V=O$ bond:

$$VOCl_3 + Al(C_2H_5)_3 \longrightarrow Cl_3V\begin{matrix}\diagup OAl(C_2H_5)_2 \\ \diagdown C_2H_5\end{matrix} \longrightarrow VCl_3 + \overset{\diagup OC_2H_5}{Al(C_2H_5)_2}$$

The trivalent and tetravalent vanadium compounds formed in excess triethylaluminum then react in a similar manner:

$$VCl_3 + Al(C_2H_5)_3 \longrightarrow C_2H_5VCl_2 + Al(C_2H_5)_2Cl$$

$$C_2H_5VCl_2 \longrightarrow VCl_2 + C_2H_5\cdot$$

$$VOCl_2 + Al(C_2H_5)_3 \longrightarrow Cl_2V\begin{matrix}\diagup OAl(C_2H_5)_2 \\ \diagdown C_2H_5\end{matrix} \longrightarrow VCl_2 + \overset{\diagup OC_2H_5}{Al(C_2H_5)_2}$$

Thus, in the formation of one mole of $Al(C_2H_5)_2Cl$ the valency of vanadium decreases by one unit, while in the formation of one mole of $Al(C_2H_5)_2OC_2H_5$ it decreases by two units. If the initial molar ratio $Al(C_2H_5)_3:VOCl_3$ is more than two, the content of chlorine in the solution increases. This is explained by the difficulty with which vanadium is reduced to V(I) and to the establishment of equilibrium.

$$VCl_2 \cdot Al(C_2H_5)_2Cl + Al(C_2H_5)_3 \rightleftharpoons VCl_2 \cdot Al(C_2H_5)_3 + Al(C_2H_5)_2Cl$$

We may assume that $VCl_2 \cdot Al(C_2H_5)_2OC_2H_5$ reacts in a similar manner:

$$VCl_2 \cdot Al(C_2H_5)_2OC_2H_5 + Al(C_2H_5)_3 \rightleftharpoons VCl_2 \cdot Al(C_2H_5)_3 + Al(C_2H_5)_2OC_2H_5$$

Shulyndin et al. /63/ published interesting results of a structural study of a complex formed by the reaction between alkylaluminum compounds and vanadium tetrachloride. They took EPR spectra of these compounds in benzene solution. Triethylaluminum and triisobutylaluminum react with VCl_4 almost instantaneously, with formation of dark-colored solutions of derivatives of bivalent vanadium, which, according to the author, are true solutions; this was noted at an earlier date by Carrick /64 — 66/. If the amount of the alkylaluminum compound is deficient, a brown precipitate separates out, owing to the formation of trivalent vanadium compounds. It was found by EPR spectroscopy that after vanadium tetrachloride has been reduced by triethylaluminum to VCl_2, the atom of bivalent vanadium remains bound to four equivalent chlorine atoms. Such a structure may be of the bridge type, for example

$$\begin{matrix} R & & Cl & & Cl & & R \\ & Al & & V & & Al & \\ R & & Cl & & Cl & & R \end{matrix}$$

Bier et al. /67/ studied the reducing tendency of organoaluminum compounds with respect to certain derivatives of vanadium. They showed that when triethylaluminum, diethylaluminum chloride or ethylaluminum dichloride are made to react with tetravalent vanadium, the vanadium is reduced to a valency of about 2.5 — 3 during the first ten minutes. The larger the number of alkyl groups in the organoaluminum compound, the stronger its reducing power. Other chlorinated compounds of pentavalent vanadium, such as $VOCl_3$, $VO(OC_2H_5)Cl_2$, $VO(OC_2H_5)_3$ or $VOCl_2(C_5H_7O_2)$, display a similar behavior towards alkylaluminum compounds, i. e., the reduction of vanadium to V(III) is very fast in all cases. Vanadium halides are reduced to V(III) even at −78°C. When VCl_4 is reduced by alkylaluminum halides, various reduction products are formed, the identities and the amounts of which depend on the alkylaluminum halide used. When VCl_4 is reduced by ethylaluminum dichloride, the formation of the voluminous precipitate, which mainly contains compounds of trivalent vanadium, is accompanied by formation of soluble reaction products, mostly consisting of bivalent vanadium compounds. The precipitate has vanadium trichloride as its major component, which still

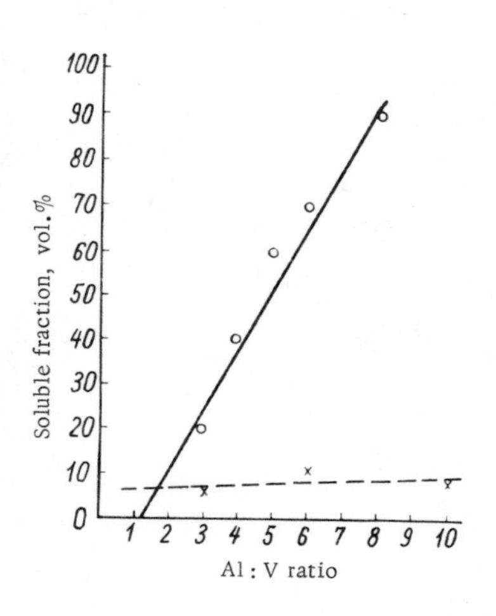

FIGURE 13. Yields of soluble and insoluble products of the reaction between VCl_4 and $C_2H_5AlCl_2$ (——) or $(C_2H_5)_2AlCl$ (---) as the function of the molar ratio Al : V (concentration of VCl_4 : 100 mmoles/liter of heptane; reaction temperature 25°C).

contains large amounts of the organoaluminum compound or of complexed $AlCl_3$; these can be only partly removed by organic solvents. The soluble fraction contains chlorinated compounds of bivalent vanadium which are complexed with alkylaluminum dichloride or with $AlCl_3$, and an excess of the organoaluminum compound. The soluble reaction products are very sensitive to oxygen and decompose above 80°C with the liberation of the green VCl_2.

It is seen from Figure 13 that the reaction between VCl_4 and $Al(C_2H_5)_2Cl$ yields much more sparingly soluble compounds of bivalent vanadium than does the reaction with $C_2H_5AlCl_2$. This would suggest that $C_2H_5AlCl_2$ is able to keep the bivalent vanadium products in solution, probably by complex formation, whereas $Al(C_2H_5)_2Cl$ and $Al(C_2H_5)_3$ have no such effect.

It was also noted that the gaseous reaction products which are evolved during the reaction between vanadium compounds and alkylaluminum compounds have a practically identical composition and are evolved in identical amounts. It was shown by Bier et al. /67/ that in the reduction of VCl_4 it is necessary to assume the same alkylation mechanism and mechanism of decomposition of the alkylvanadium halide as for the case of the reduction of $TiCl_4$ by organoaluminum compounds. The effect of electron donors on the reduction of VCl_3 by alkylaluminum compounds was also studied /67a/.

Nasirov et al. /68/ studied the reaction between vanadyl acetylacetonate $VO(C_5H_7O_2)_2$ and triethylaluminum, the products of which effectively catalyze the polymerization of acetylenic compounds. The reaction was conducted in benzene by mixing triethylaluminum and $VO(C_5H_7O_2)_2$ in appropriate amounts. The fully soluble reaction product — the brown-colored organometallic complex, with molar ratio Al : V = 4 : 1 — was studied by EPR spectroscopy and by determination of magnetic susceptibility. The experimental results showed that the vanadium in vanadyl acetylacetonate is reduced as follows:

$$\underset{\underset{O}{\|}}{V^{4+}}\left[\begin{matrix} O{-}C(CH_3) \\ \quad \| \\ \quad CH \\ O{=}C(CH_3) \end{matrix}\right]_2 + 2Al(C_2H_5)_3 \text{—} \underset{\underset{O}{\|}}{V}\left[\begin{matrix} O{-}C(CH_3) \\ \quad \| \\ \quad CH \\ O{=}C(CH_3) \\ \downarrow \\ 2Al(C_2H_5)_3 \end{matrix}\right]_2 \longrightarrow$$

$$\longrightarrow \underset{\underset{O}{\|}}{V^{2+}}\left[\begin{matrix} O{-}C(CH_3) \\ \quad CH \\ O{-}C(CH_3) \\ | \\ Al(C_2H_5)_2 \end{matrix}\right]_2 + C_2H_5$$

(I)

$$\text{(I)} + 2Al(C_2H_5)_3 \longrightarrow \underset{\underset{O}{\|}}{V^{2+}}\left[\begin{matrix} 2Al(C_2H_5)_3 \\ \uparrow \\ O{=}C(CH_3) \\ \quad CH \\ O{-}C(CH_3) \\ | \\ Al(C_2H_5)_2 \end{matrix}\right]_2$$

(II)

Subsequently /69/ the authors came to the conclusion that the last stage does not result in the formation of compound (II) alone, but also in the alkylation of the vanadium compound:

$$(I) + 2Al(C_2H_5)_3 \longrightarrow \underset{C_2H_5}{\overset{V^{2+}}{|}} \left[\begin{array}{c} \cdot O{=}C(CH_3) \\ CH \\ \cdot O{=}C(CH_3) \\ | \\ Al(C_2H_5)_2 \end{array} \right]_2 + (C_2H_5)_2Al{-}O{-}Al(C_2H_5)_2$$

(III)

According to the authors, complex (III) is the active agent in the polymerization of acetylenic compounds. Nasirov also showed that the reaction between vanadyl acetylacetonate and $Al(C_2H_5)_2Cl$ results only in a partial reduction of vanadium, whereas $Al(C_2H_5)Cl_2$ does not reduce the vanadium in the compound at all.

The reactions between tetraphenyltin, aluminum tribromide and vanadium tetrachloride, and the composition of the catalyst complex constituted by these compounds, were studied by a number of workers. This was done in order to clarify the catalytic mechanism of this complex which was first proposed in a modified form by Carrick /65, 66, 70/. One of the most important studies on the subject is that by Stotskaya and Krentsel' /70/. Their experimental results show that the interaction between the components does not begin with the arylation of vanadium chloride by tetraphenyltin to the organovanadium compound of the type RVX, but rather with the arylation of the aluminum bromide:

$$Sn(C_6H_5)_4 + 2AlBr_3 \longrightarrow 2Al(C_6H_5)Br_2 + Sn(C_6H_5)_2Br_2$$

The resulting phenylaluminum dibromide then reacts with vanadium tetrachloride with formation of the soluble complex catalyst, as is also the case in conventional organometallic catalysts. The solubility of these products is due to the weak reducing power of $Al(C_6H_5)Br_2$ with respect to vanadium tetrachloride. The authors obtained a direct confirmation of the structure and composition of the intermediate compounds produced in the course of formation of the complex catalyst by studying the reaction with the use of Mössbauer effect. The formation of the organometallic active complex from phenylaluminum dibromide and titanium tetrachloride involves the following intermediate stage:

$$VCl_4 + Al(C_6H_5)Br_2 \longrightarrow Cl_3V{-}Cl\cdots Al(C_6H_5)Br_2$$

Subsequently, in the presence of ethylene the catalyst complex assumes the following structure:

$$\underset{\substack{\uparrow \\ CH_2=CH_2}}{Cl_3V}-Cl-Al(C_6H_5)Br_2 \longrightarrow \left[Cl_3V\begin{matrix} \diagup CH_2-CH_2 \diagdown \\ \diagdown Cl \text{———} Al(Br)_2 \end{matrix} C_6H_5 \right] \longrightarrow \underset{\substack{| \\ CH_2=CH_2}}{Cl_3V}(CH_2-CH_2-C_6H_5)-Cl-AlBr_2$$

transition compound

active complex

The entry of the first monomer molecule and the formation of the active complex represent the initiation step of the polymerization.

Bibliography

1. Gaylord, N.G. and H.F. Mark. Linear and Stereoregular Addition Polymers. — New York, Interscience. 1959.
2. Chem. Ind. Ltd; Belgian Patent 543941. 1956.
3. Ziegler, K. — Angew. Chem., **68**:581. 1956.
4. Furukawa, J. et al. — J. Polymer Sci., **28**(117):450. 1958.
5. Roha, M. et al. Papers Presented at the Boston Meeting, Division of Paint, Plastics and Printing Ink Chemistry, **19**(1):120. 1959.
6. French Patent 1146245. 1957.
7. Berger, M. and T. Boultber. — J. Appl. Chem., **9**(9):490. 1959.
8. BASF, US Patent 2905661. 1959.
9. Brebner, D. et al. — US Patent 2822357. 1958.
10. Beili, F. et al. — US Patent 2912425. 1959.
11. Natta, G. et al. — Gazz. chim. ital., **87**:549. 1957.
12. Malatesta, A. — Can. J. Chem., **37**(7):1176. 1959.
13. Simon, A. et al. — Monatsh. Chem., **90**(4):443. 1959.
14. Copper, M. and J. Rose. — J. Chem. Soc., No. 2:795. 1959.
15. Boldyreva, I.I., B.A. Dolgoplotsk, and V.A. Krol'. — VMS, **1**(6):900. 1960.
16. Kern, R. and H. Hurst. — Polymer Sci., **44**(143):272. 1960.
17. Pozamantir, A.G., A.A. Korotkov, and I.S. Lishanskii. — VMS, **1**(8):1207. 1959.
18. Friedländer, H. and K. Oita. — Ind. Eng. Chem., **49**:1885. 1957.
19. Kovacs, L., A. Simon, and D. Gimesch. — Magyar Kémikusok lapja, **13**:180. 1958.
20. Rodriguez, L. et al. — Tetrahedron Letters, No. 17:7. 1959.
21. Ziegler, K. — Brennstoff Chem., **35**:321. 1954.
22. Ziegler, K. — Angew. Chem., **67**:541. 1956.
23. Simon, A. et al. — Magyar Kémiai Folyóirat, No. 6. 1959.
24. Adema, E., H. Bartelink, and J. Smidt. — Rec. trav. chim. Pays-Bas, **80**:173. 1961.
25. Adema, E., H. Bartelink, and J. Smidt. — Rec. trav. chim. Pays-Bas, **81**:73. 1962.
26. Varodi, E., V.I. Tsvetkova, and N.M. Chirkov. — Doklady AN SSSR, **152**:908. 1963.
27. Groenewege, M. — Z. phys. chem. (Frankfurt), **18**:147. 1958.
28. Gray, A. — Can. J. Chem., **41**:1511. 1963.

29. Gray, A. et al. — Can. J. Chem., **41**:1502. 1963.
30. Beermann, C. and H. Bestian. — Angew. Chem., **71**:918. 1959.
31. Jones, M. et al. — Can. J. Chem., **38**:2303. 1960.
32. Sakurada, A. et al. — J. Phys. Chem., **68**(7):1934. 1964.
32a. Bohar, D. et al. — J. Organomet. Chem., **4**:278. 1965.
33. Natta, G. — Actes du 2 Congrès Internat. Catalyse, Paris. 1960.
34. Natta, G. — Chim. e ind., **42**:1207. 1960.
35. Feay, C. — US Patent 3274223. 1966.
36. Rodriguez, L. and G. Gabant. — J. Polymer Sci., **57**:883. 1962.
37. Fontana, G. and R. Osborne. — J. Polymer Sci., **47**:522. 1960.
38. Natta, G. — J. Polymer Sci., **34**:21. 1959.
39. Simon, A., L. Kollac, and L. Koröck. — Monatsh. Chem., **95**(3): 842. 1964.
39a. Meizlik, J. et al. — Chem. průmysl, **15**(2):85. 1965.
39b. Ambroz, J. et al. — Chem. průmysl, **17**(2):69. 1967.
40. Natta, G. et al. — J. Am. Chem. Soc., **79**:2975. 1957.
41. Natta, G. et al. — J. Am. Chem. Soc., **80**:755. 1958.
41a. Corradini, P. and A. Sirigu. — Inorg. Chem., **6**(3):601. 1967.
42. Breslow, D. and N. Newburg. — J. Am. Chem. Soc., **79**:5072. 1957.
43. Breslow, D. and N. Newburg. — J. Am. Chem. Soc., **81**:81. 1959.
44. Long, W. and D. Breslow. — J. Am. Chem. Soc., **82**:1955. 1960.
45. Kissin, Yu. V., E. V. Tolstykh, and N. M. Chirkov. — Doklady AN SSSR, **145**(1):104. 1962.
46. Shilov, A. E., A. K. Zefirova, and N. N. Tikhomirova. — ZhFKh, **33**(9):2113. 1959.
47. Zefirova, A. K., N. N. Tikhomirova, and A. E. Shilov. — Doklady AN SSSR, **132**(5):1082. 1960.
48. Zefirova, A. K. and A. E. Shilov. — Doklady AN SSSR, **136**(3):599. 1961.
49. Shilov, A. E., A. K. Shilova, and B. N. Bobkov. — Vysokomolekulyarnye Soedineniya, **4**:1688. 1962.
50. Maki, A. and E. Randall. — J. Am. Chem. Soc., **82**:4109. 1960.
51. Stepovik, L. P., A. K. Shilova, and A. E. Shilov. — Doklady AN SSSR, **148**(1):122. 1963.
51a. Grigoryan, E. A., F. S. D'yachkovskii, G. M. Khvostik, and A. V. Shilov. — Vysokomolekulyarnye Soedineniya, **A9**(6):1233. 1967.
52. Bartelink, H., H. Bos, and J. Smidt. — Congrès Ampère, Leipzig, 13—16 Sept. 1961.
53. D'yachkovskii, F. S., P. Ya. Yarovitskii, and V. F. Bystrov. — Vysokomolekulyarnye Soedineniya, **6**(4):659. 1964.
54. Grigoryan, E. G., F. S. D'yachkovskii, and A. E. Shilov. — VMS, **7**(1):145. 1965.
54a. D'yachkovskii, F. S. and A. E. Shilov. — ZhFKh, **41**(10):2515. 1967.
55. Fushman, E. A., V. I. Tsvetkova, and N. M. Chirkov. — Doklady AN SSSR, **164**(5):1085. 1965.
56. Natta, G. and G. Mazzanti. — Chim. e ind., **39**(9):733. 1957.
57. Mazzanti, G. et al. — Chim. e ind., **39**(9):743. 1957.
58. Mazzanti, G. et al. — Chim. e ind., **39**(10):825. 1957.
59. Natta, G. et al. — Chim. e ind., **40**(5):362. 1958.
60. Natta, G. et al. — Chim e ind., **41**(2):116. 1959.

61. Natta, G. et al. — Rend. Accad. Naz. Lincei, **24**(5):479. 1958.
62. Petrov, G. N. and A. A. Korotkov. — Doklady AN SSSR, **141**(3):632. 1961.
63. Shulyndin, S.V. et al. — Zhurnal Strukturnoi Khimii, **2**(6):740. 1961.
64. Carrick, W. — J. Am. Chem. Soc., **80**:6455. 1958.
65. Carrick, W. — J. Am. Chem. Soc., **82**:3883. 1960.
66. Carrick, W. — J. Am. Chem. Soc., **82**:5319. 1960.
67. Bier, G., A. Gumboldt, and G. Schleitzer. — Makromol. Chem., **58**:43. 1962.
67a. Cooper, W. et al. — J. Polymer Sci., Pt. B., **4**(5):309. 1966.
68. Nasirov, F. M., G. P. Karpacheva, B. E. Davydov, and B. A. Krentsel'. — Izvestiya AN SSSR, Seriya Khimicheskaya, **9**:1697. 1964.
69. Nasirov, F. M. Author's Summary of Thesis. INKhS AN SSSR, Moskva. 1965.
70. Stotskaya, L. L. Author's Summary of Thesis. INKhS AN SSSR, Moskva. 1964.

Chapter IV

METHODS USED IN THE ANALYSIS OF COMPONENTS OF COMPLEX CATALYSTS

The necessity for a very accurate determination of the contents of both the organoaluminum compounds and the transition metal salts is a direct consequence of the high reactivities of the components forming part of the catalyst complex and of their apparently unchanged external aspect when acted upon by oxygen, water, sulfurous compounds and other compounds present in the air and in solvents.

Despite the abundance of literature publications on the analysis of these compounds, there is as yet no general agreement in favor of any particular method for the determination of the complex components. In our view this lack of agreement is due to the fact that methods which would be sufficiently accurate, sensitive, rapid, and technically simple have not yet been proposed.

This chapter will deal with the most frequently used analytical methods for the determination of organoaluminum compounds and halides of a number of transition metals.

I. ANALYSIS OF ORGANOALUMINUM COMPOUNDS

The methods for the analysis of organoaluminum compounds, proposed during the past decade, include the following element determinations /1 — 20/: 1) aluminum in alkylaluminum compounds and their derivatives /21 — 53/; 2) gas-volumetric determination of the hydrocarbon part, with subsequent analysis of the composition of the liberated gases /32, 57 — 63/; 3) determination of alkoxy groups in partly oxidized alkylaluminum compounds /64 — 73/; 4) iodometric determination of $>$Al — R bond /52, 54, 55/. In addition, many publications deal with the determination of the so-called active part of the alkylaluminum compound R_2AlX, where X = R, H, Hal. They include several varieties of electrometric titration /59, 86 — 90/, the ammonia method /85/ and determinations of the concentration of alkylaluminum compounds by reduction of $TiCl_4$ /74 — 82/, titration of the dielectric constant /91 — 95/, calorimetric titration /96, 97/, determinations based on reversible color reactions (indicator methods) /32, 59, 99, 100/ and spectrophotometric titration /61, 98, 100 — 102/.

Methods used in elementary analysis of organoaluminum compounds, functional group assay and determination of the active component in these compounds are described below. The most important methods for the determination of alkylaluminum compounds are also described.

1. DETERMINATION OF THE ELEMENTARY COMPOSITION AND FUNCTIONAL GROUP ASSAY IN ALKYLALUMINUM COMPOUNDS

Elementary analysis

The determination of the elementary composition of organoaluminum compounds involves the decomposition of the compound in some suitable manner, followed by a quantitative analysis of the resulting products. Depending on the mode of decomposition, the elementary analysis may be effected by dry oxidation or by hydrolytic decomposition of the sample. In either case, the sample taken will be readily decomposed in the air; in fact, the necessity for sampling and weighing in an atmosphere of inert gas is one of the main difficulties which are encountered in the analysis of alkylaluminum compounds.

Many sampling techniques have been proposed /1 — 8/, but none is simple enough for rapid sampling of very small volumes of alkylaluminum compounds to be possible. The sampling is conducted in special airtight vessels in which an inert gas atmosphere is produced (nitrogen or argon); the gas must be purified from oxygen and moisture to a content of not more than 0.0005 vol.% /9/. Such vessels are most often used for the sampling of solids.

Gel'man and Bryushkova /1/ described a technique for the sampling of liquids spontaneously flammable in the air, which was used with success in the analysis of alkylaluminum compounds and dialkylaluminum halides. The weighed sample is placed in a capillary. The elementary analysis of alkylaluminum compounds thus sampled is complicated by the fact that the capillary with the sample must not be sealed in order to avoid a partial decomposition of the product. Accordingly, the sample must be weighed in an open capillary immediately before the combustion. It was experimentally established that the weight of the capillary with the sample remains unchanged for a few minutes. It is preferable to use quartz glass capillaries, since these can be re-used.

The authors developed a technique for the sampling of alkylaluminum compounds in open quartz glass capillaries with the aid of the instrument represented in Figure 14. Prior to sampling, purified argon is passed through the system for 15 minutes, and the weighed capillary is introduced without interrupting the stream of argon. After a part of the capillary has been filled, the capillary is held over the liquid so as to fill its lower part with argon; in this way the sample is in contact with argon on both sides of the capillary. The capillary is then taken out, its tip is carefully wiped with a chamois cloth and weighed. The usual sample size is 3 — 12 mg.

An even better procedure /2/ for sampling alkylaluminum compounds and their solutions is to weigh the capillary filled with the inert gas and to introduce it in a stream of nitrogen into the filling unit represented in Figure 15. The capillary, suspended from a Perlon thread, is lowered by turning the ground glass joint so that its lower tip dips into the liquid to be analyzed. The nitrogen is now very cautiously removed so as to produce a low vacuum in the instrument; when nitrogen is readmitted, a sufficient amount of the liquid will enter the capillary. The capillary is then held above the liquid surface and the liquid rises in the capillary. Thus both the lower and the upper surface of the liquid in the capillary are in contact with nitrogen. The capillary is taken out, carefully wiped with chamois cloth and weighed.

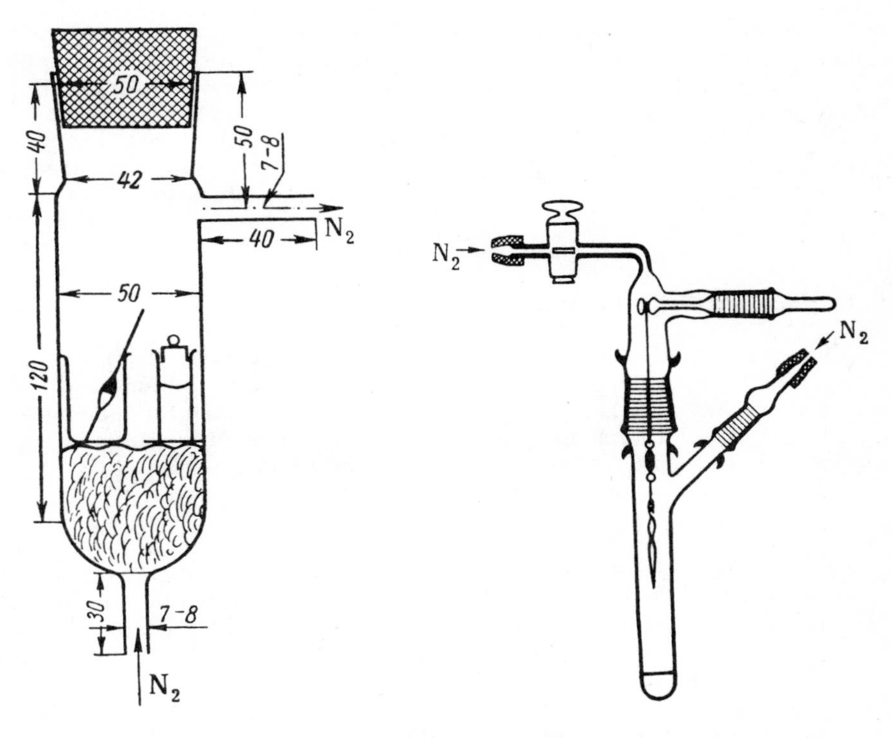

FIGURE 14. Sampling organoaluminum compounds into open quartz glass capillaries.

FIGURE 15. Filling the capillary with the organoaluminum compound.

In semimicro determinations, alkylaluminum compounds may be sampled into special ampules. Spherical glass ampules, 10 — 12 mm in diameter, with a bent capillary, are weighed, placed in a special tube and evacuated, after which the tube is filled with argon. The evacuation and the filling with inert gas is repeated twice. The ampule is taken out of the tube and its capillary is rapidly lowered into the vessel with the liquid by passing it through a bore in a cork. The borehole is rinsed with a stream of inert gas fed from a rubber tube. A very small Dewar flask with acetone-dry ice freezing mixture is set under the globe of the ampule, and when the ampule is about half-filled, its tip is lifted above the liquid level. The ampule is then rapidly sealed and weighed. For the concentrated product the sample size should be 0.06 — 0.15 gram.

The contents of carbon, hydrogen, aluminum and halogens in alkylaluminum compounds are determined by igniting the sample in oxygen at a high temperature. This method is widely employed in determining the composition of alkylaluminum compounds, in particular the variant of Gel'man and Bryushkova /1/, in which all elements are determined on one sample. Carbon and hydrogen are determined by pyrolysis in an evacuated quartz glass tube.

The combustion is effected using the conventional apparatus and absorber system /10, 11/. Aluminum is determined by weighing as Al_2O_3; the result must be corrected for the hygroscopicity of alumina in the given tube and for the difference between the weight of the tube when filled with oxygen and when filled with air. The halogen content is found by determining the increase in weight of a silver-containing quartz cartridge. The accuracy of carbon and hydrogen determinations is that usually encountered in elementary analysis. The permissible error in the determination of aluminum and halogens is 0.4 — 0.5% (abs.). The determination takes 25 — 30 minutes.

Determination of total aluminum

The determination of the total aluminum in organoaluminum compounds is effected by hydrolysis of the alkylaluminum sample, which is followed by the determination of the aluminum ion. The sample may be decomposed by hydrolyzing agents such as water, alcohol, aqueous acid solutions, ethylene glycol, etc. /21 — 32/. The reaction takes place as follows:

$$AlR_3 + 3H_2O \longrightarrow Al(OH)_3 + 3RH$$

Thus, the determination of the total aluminum in organoaluminum compounds (after decomposition) consists in determining the aluminum in the inorganic compound by one of the methods described below.

Grosse and Mavity /21/ dissolved the aluminum hydroxide formed by the above reaction in acid and precipitated the aluminum by 8-hydroxyquinoline; this was followed by gravimetric or volumetric determination. Ziegler et al. /32/ initially converted aluminum hydroxide to fluoroaluminate, the latter being subsequently titrated as the salt of a weak acid. Alternatively, the hydrolyzed sample may be treated with hydrochloric, sulfuric or nitric acid, after which the aluminum oxide is ignited in a platinum crucible to constant weight /2, 33/.

Methods for the determination of aluminum in the presence of other elements have been developed. Boldyreva et al. determined Al in the presence of Ti by treating a sample of hydrolyzed alkylaluminum compound dissolved in sulfuric acid with cupferron. Following the extraction of the cupferronate with chloroform the solution was neutralized with dry sodium acetate, aluminum was precipitated from the neutral solution by 8-hydroxyquinoline and ignited to Al_2O_3 at 1000 — 1100°C /34/. Burke and Davis /35/ developed a rapid, accurate titrimetric method for the determination of Al in the presence of Ni, Fe and Cu by 1, 2-cyclohexylenedinitrilotetraacetic acid. The absolute error did not exceed 0.08% if the aluminum content was between 0.5 and 93.1% /35/. Babenyshev and Kuznetsova described an amperometric determination of aluminum ions in aqueous solutions by

Trilon B, without separating aluminum from the other components /36/. The accuracy was 0.2% (abs.). Mikhailenko /37/ proposed an accurate method for the determination of aluminum ion by p-hydroxy-m-nitrophenylphosphinic acid. A spectrophotometric determination of aluminum by Stilbazo has also been described /38/. Stilbazo forms a colored complex with aluminum ion, which absorbs at 500 mμ. The color is stable for 3 hours; the optimum pH is 5.65.

The complexometric determination of aluminum has been the subject of numerous studies. The reagent employed is a standard solution of the sodium salt of ethylenediamine tetraacetic acid, also known as Trilon B, Complexone III or Chelatone III. Aluminum slowly reacts with Trilon B at room temperature in a wide pH range (between 1.5 and 12). When aluminum is determined by complexometric back-titration /22 — 30, 39 — 51/, a definite volume of standard solution of Trilon B is added to the sample solution, and the excess is determined by titration against a solution of a suitable metal salt.

Direct titration of aluminum against Trilon B cannot be carried out, since the indicator is blocked by aluminum cations, with which it forms a very stable red-violet complex, which does not react with Trilon B (for this reason there is no color change at the equivalence point) /27, 28, 52/. According to other workers /39 — 43/ direct titration is not possible because the indicators (Alizarin, Stilbazo, hematoxylin, Chromeazurol S and others), which form colored compounds with aluminum, are decomposed by Trilon B.

The pH of the medium must be controlled; at pH 4 and above hydrolysis takes place and some of the aluminum does not become bound by Trilon during the titration; in such a case good results are obtained by indirect determination in the presence of various indicators /28/. If the solution is back-titrated, the unbound Trilon in excess may be determined at different pH values /39/. In order to prevent the reaction with hydrolyzed aluminum ions or with aluminum hydroxide, Trilon must be introduced into acid (approximate pH range 1 — 2) or into alkaline (approximate pH range 12 — 13) solutions of aluminum /39, 41, 43/. Subsequently, the solutions are heated to boiling and neutralized to the desired pH value while hot. If this is done, the trilonate will be formed rapidly and quantitatively during the neutralization itself, which should be effected with the aid of weak acids and bases. The appropriate buffer solution is then added to the sample. A number of workers have recommended that Trilon B be introduced into solutions of aluminum at pH 3.6 — 4.3 and that they then be boiled /44/. It is also possible to introduce Trilon B into a cold, acid solution of an aluminum salt, neutralize and heat to boiling /45 — 47/. Under these conditions hydrolysis may take place and the amount of the Trilon consumed is not equivalent to the content of aluminum /39/. During the back-titration, the cation of the titrant solution must not decompose the aluminum trilonate formed /48/.

Přibil et al. /49/ use a solution of zinc salt as titrant; the titration is conducted at pH 10, when zinc trilonate is more stable than aluminum trilonate. The zinc ions decompose the aluminum trilonate formed, and the equivalence point is difficult to determine /39/. Some workers /46, 50/ titrate with solutions of zinc salts at pH 5 — 7, using redox indicators in the presence of ferric ferrocyanide or Eriochrome Cyanine R. This technique is more suitable, since at pH 5 — 7 aluminum and zinc trilonate are about equally stable /39/.

If the excess Trilon in solution is titrated at pH 6 against an acid (pH 1.5 — 2) standard solution of aluminum ion itself, the previously formed aluminum trilonate cannot be decomposed. The titrant aluminum ions are not hydrolyzed under these conditions /39/. The method was proposed by Bashkirtseva and Yakimets /41/ to determine aluminum in a number of materials.

Titration of the excess Trilon at pH 6 against a solution of ferric iron has the advantages of the ready availability of the reagents and of a sharp end point. At this pH value, ferric trilonate is less stable than aluminum trilonate. Salicylic acid /45/ and sodium sulfosalicylate /43, 51/ have been proposed as indicators in this titration.

Complexometric determinations of aluminum in the presence of various indicators have been described /46/: Eriochrome Cyanine R, disubstituted sodium phosphate, ferroferricyanobenzidine, ferroferricyanodimethyl-naphthidine. Complexometric titration of aluminum in the presence of titanium has been described /53/. The method was checked on samples having different concentrations and different Al:Ti ratios.

The trilonometric method is applicable within the concentration range of 1 — 500 mg Al per liter. Calcium, magnesium, cupric, zinc and manganous ions interfere.

All these methods for the determination of aluminum have a variable accuracy, not exceeding ± 5.0 rel. %.

In our view, the best one of the methods just described is to complex aluminum ion with Trilon B in acid solution and to titrate the excess Trilon B against a standard solution of zinc chloride.

Functional group assay

Quantitative determination of the functional groups (R, H, OR and other groups bound to Al) in organoaluminum compounds is based on their hydrolytic cleavage. The alkyl and the hydride groups bound to aluminum are then converted to alkanes and to hydrogen, while the alkoxyl groups are converted to alcohols.

Alkyl, alkenyl, hydride and alkoxy groups directly bound to the aluminum can be determined by methods based on the hydrolytic cleavage of these groups.

Hydroxyl groups bound to aluminum can be determined by the reaction between alkylaluminum compounds and lithium aluminum hydride /56/.

Determination of alkyl groups

Gas-volumetric method. This is a method for the determination of alkyl groups in lower alkylaluminum compounds, such as trimethylaluminum, triethylaluminum, tripropylaluminum and tributylaluminum. The alkyl groups are cleaved by hydrolyzing agents, the volume of the liberated gas is measured and the composition is subsequently determined. Thus, for instance, in the alcoholysis

$$AlR_3 + 3R'OH \longrightarrow Al(OR')_3 + 3RH$$
$$AlR_2H + 3R'OH \longrightarrow Al(OR')_3 + 2RH + H_2$$

three moles of gas are liberated by one mole of AlR_3. The hydrolysis of halogen-substituted or alkoxy-substituted aluminum compounds results in the formation of the corresponding acid or alcohol:

$$R_2AlX + 3R'OH \longrightarrow Al(OR')_3 + 2RH + HX$$
$$R_2AlOR + 3R'OH \longrightarrow Al(OR')_3 + 2RH + ROH$$

which dissolves in the reagent and does not interfere with the determination of alkyl and hydride groups /57/. When alkylaluminum hydrides are decomposed, the contents of hydrogen and of the hydrocarbons in the gas are determined by conventional chromatographic methods /32, 57, 58/. According to some workers, the hydrolysis of alkylaluminum compounds yields saturated hydrocarbons only, in conformity with the above equations /21/. According to others, decomposition by alcohols is accompanied by the formation of unsaturated hydrocarbons, which means that side reactions take place /59/.

Alcoholysis gives good results in the analysis of triethylaluminum derivatives only, but trialkylaluminum compounds, beginning with tripropylaluminum, give low results, since the evolved alkane is partly soluble in alcohol. In analyzing such compounds, water only should be used, or else the product should be heated following the decomposition /60/.

Bonitz /59/ and Ziegler et al. /32/ used the gasometric method in the analysis of triethylaluminum, tripropylaluminum, triisobutylaluminum and similar low-molecular compounds, which were decomposed by the high-boiling alcohol 2-ethylhexanol. The alcoholysis of each alkyl group yielded 1 mole of alkane, while the alcoholysis of each hydride group yielded one mole of hydrogen. The contents of the alkyl and hydride groups in the samples were calculated from the amount of the gas evolved. Ziegler used mass-spectrometric gas analysis to determine the composition of the liberated gas mixture /32/. Crompton et al. /57/ studied the alcoholysis of alkylaluminum compounds and recommended the most suitable reagents for the quantitative decomposition of each type of organoaluminum compound, viz., for triethylaluminum a 4:1 mixture of hexane and monoethylene glycol; for the determination of tributylaluminum, tripropylaluminum and of hydrogen and lower alkyls in higher alkylaluminum compounds a 3:1 mixture of monoethylene glycol with 20% sulfuric acid.

Neumann /61/ developed a method for the determination of the number of Al — H bonds in the presence of other alkylaluminum compounds, based on the reaction of AlR_3, R_2AlH, R_2AlX and R_2AlOR with secondary amines such as N-methylaniline. The experimental conditions are so chosen that the secondary amine reacts exclusively with the dialkylaluminum hydride. In this way the content of dialkylaluminum hydride could be determined in the presence of trialkylaluminum. The compounds R_2AlX and R_2AlOR are less reactive than AlR_3, and do not interfere with the determination. Compounds such as $RAl(OR)_2$ and $RAlX_2$ must not be present together with trialkylaluminum compounds, since they react with them, forming the corresponding R_2AlOR or R_2AlX. The accuracy of an individual determination is to within ± 5.0 rel. %.

Lionozova and Genusov /62/ reported the results of the determination of alkyl groups in hydrocarbon solutions of organoaluminum compounds and showed that it is necessary to correct for the solubility of the evolved alkane. A simple and rapid method for the determination of alkyl groups was proposed by Shvindlerman and Zavadovskaya /63/, who decomposed alkylaluminum compounds with nitric acid at a low temperature and measured the volume of the gas evolved. Gas-volumetric methods give a good idea of the composition of the substance, but are too time-consuming to be used in industry.

Iodometric method. It was shown by Bartkiewicz and Robinson /54/ that triethylaluminum reacts with iodine according to the equation:

$$Al(C_2H_5)_3 + 3I_2 \longrightarrow AlI_3 + 3C_2H_5I$$

The reaction is quantitative. In the presence of other organoaluminum compounds, e. g., diethylaluminum chloride or diethylaluminum ethoxide, the reaction also takes place with the rupture of an Al — C bond and formation of the corresponding aluminum compounds.

$$Al(C_2H_5)_2Cl + 2I_2 \longrightarrow AlI_2Cl + 2C_2H_5I$$

$$Al(C_2H_5)_2OC_2H_5 + 2I_2 \longrightarrow AlI_2OC_2H_5 + 2C_2H_5I$$

The authors suggested that this reaction be used in quantitative determination of organoaluminum compounds. Crompton /55/ subsequently studied the reaction in detail and developed a method for the determination of organoaluminum compounds in hydrocarbon solutions. According to this worker, the solution of the organoaluminum compound must be introduced drop by drop into a solution of iodine (at least 30% excess) in order to avoid side reactions and to obtain reproducible results. The reaction between iodine and dialkylaluminum alkoxide takes place in different manners, depending on the experimental conditions. It was noted that the standard errors in the determination of the iodine number of three types of organoaluminum compounds — trialkylaluminum compounds, dialkylaluminum alkoxides and dialkylaluminum chlorides — are relatively low. The author reported comparative results of determinations of triethylaluminum and tripropylaluminum by three different methods: by determination of organically bound aluminum, by the isoquinoline method and by the iodometric method.

Solomon et al. /52/ conducted a statistical analysis of the most important methods for the determination of organoaluminum compounds and pointed out that the iodometric method is suitable for the determination of alkylaluminum compounds in dilute solutions. The iodometric method also serves in the studies of the alterations in the structure of hydrocarbon solutions of alkylaluminum compounds on storage. It is applicable to solutions of trialkylaluminum compounds, dialkylaluminum chlorides and dialkylaluminum alkoxides, and also to high-molecular trialkylaluminum compounds.

Determination of alkoxy groups

Organoaluminum compounds, especially trialkylaluminum compounds and dialkylaluminum hydrides, even after purification, usually contain small quantities of alkylaluminum alkoxides, which are formed by the reaction between the compounds and atmospheric oxygen or the oxygen residual in the inert gas. Methods of analysis of alkylaluminum alkoxides are based on their hydrolytic cleavage, followed by the determination of the liberated alcohols according to the reaction:

$$R_2AlOR + 3H_2O \longrightarrow ROH + 2RH + Al(OH)_3$$

The small amount of alcohol which is formed as a result of the reaction is usually determined by one of two methods. In one such method, the dyestuff formed by the interaction of the alcohol with sulfanilic acid and naphthylamine /64/, or the complex of the alcohol with nitric acid solution of cerium ammonium nitrate /64 — 68/ is colorimetrically determined. In another method, the alcohol is oxidized by potassium dichromate and the excess dichromate is then back-titrated against a standard solution of $(NH_4)_2Fe(SO_4)_2$ /69/ by the electrometric technique; alternatively, it may be back-titrated by sodium thiosulfate /70 — 73/:

$$(C_2H_5O)Al(C_2H_5)_2 + 3H_2O \longrightarrow C_2H_5OH + Al(OH)_3 + 2C_2H_6$$

$$3C_2H_5OH + 2K_2Cr_2O_7 + 8H_2SO_4 \longrightarrow 2Cr_2(SO_4)_3 + 3CH_3COOH + 2K_2SO_4 + 11H_2O$$

$$K_2Cr_2O_7 + 6KI + 7H_2SO_4 \longrightarrow 3I_2 + Cr_2(SO_4)_3 + 4K_2SO_4 + 7H_2O$$

$$I_2 + 2Na_2S_2O_3 \longrightarrow 2NaI + Na_2S_4O_6$$

The main difficulty in the determination of the alkoxy groups in organoaluminum compounds is the choice of the experimental technique which would prevent the contact of the sample with atmospheric oxygen during sampling and hydrolysis. Bondarevskaya et al. /70/ developed a technique for the decomposition of alkylaluminum compounds with water and isolation of the aqueous alcohol solution for subsequent determination of the alcohol. Crompton recommended that the decomposition be effected under mild conditions, e.g., by monoethylene glycol, in order that the alcohols formed might be quantitatively trapped /65/. It was found that the oxidation of the alcohol by dichromate in sulfuric acid will determine ethanol to within $\pm$ 0.5 rel.%, and is also suitable for the determination of ethylaluminum ethoxides /72/. In the analysis of butanol, formed by hydrolysis of the corresponding alkylaluminum alkoxide, the concentration of the sulfuric acid is a major factor which determines the value of the oxidation coefficient /71/. It has also been shown that, depending on the experimental conditions employed, the alcohols can be oxidized to butyric, acetic, or formic acids, and even to CO_2. It is accordingly preferable to determine the hydroxy derivatives in butylaluminum and isobutylaluminum compounds by the colorimetric method /64 — 68/.

2. DETERMINATION OF THE ACTIVITY OF ORGANOALUMINUM COMPOUNDS

Active organoaluminum compounds are defined as pure trialkylaluminum compounds, dialkylaluminum hydrides and dialkylaluminum chlorides. Many important reactions are given exclusively by these compounds and not by compounds of the type R_2AlX, where X = OR, $OAlR_2$, SR_2, NR_2, etc. Thus, the determination of total aluminum in alkylaluminum compounds alone is an inadequate description of the product, since the latter — as a result of being kept in vessels which are insufficiently airtight — may contain products of slow oxidation and hydrolysis, such as dialkylaluminum alkoxides and alkylalumoxanes, which reduce its reactivity. The determination of the true activity of alkylaluminum compounds is therefore the most important analytical task. Several methods for the determination of the activity of organoaluminum compounds have been recently developed; they will be discussed below.

Chemical methods for the determination of concentration of alkylaluminum compounds

Reduction of titanium tetrachloride

Trialkylaluminum compounds are capable of reducing $TiCl_4$ to derivatives of lower valencies /74 — 82, 84/. The reaction takes place as follows /74 — 76/:

$$AlR_3 + TiCl_4 \longrightarrow RTiCl_3 + R_2AlCl$$

$$RTiCl_3 \longrightarrow TiCl_3 + R$$

The $TiCl_3$ thus formed is decomposed by sulfuric acid and trivalent titanium is determined, after which the amount of the trialkylaluminum compound is calculated /75, 77, 78/. Tepenitsyna and Farberov /75/ noted that the degree of reduction of $TiCl_4$ by triethylaluminum depends on the molar ratio $AlR_3:TiCl_4$. If the molar ratio is unity, the reduction proceeds only to the trivalent titanium stage. Thus, if the ratio $AlR_3:TiCl_4$ is unity or less, the amount of Ti(III) is equivalent to the amount of AlR_3 taken, as confirmed by experimental data. The authors showed /75/ that if the reduction is conducted within 3 — 5 minutes, the amount of the reduced $TiCl_3$ will accurately correspond to the amount of AlR_3 taken. Alkylaluminum alkoxides do not form complexes with $TiCl_4$. The determination of the activity of trialkylaluminum compounds from the reduced $TiCl_4$ is simple and the titration results are accurate. It was pointed out by Habvinga and de Jong /78/ that the error is less than 10 rel. %. A determination takes $1\,^1/_2$ — 2 hours. It was shown by Solomon et al. /52/ that the most suitable method for the determination of the activity of alkylaluminum compounds is the reduction of titanium tetrachloride. The results are reproducible, accurate and no complicated apparatus is required. These workers compared the results of determination of the activity of triethylaluminum by titanium tetrachloride, by potentiometric isoquinoline titration and by determining the total aluminum content. The first two methods, which measure the activity, gave similar results, but the results of the determination of total aluminum were discrepant.

Ammonolysis /85/

This method is suitable for the determination of trialkylaluminum compounds and dialkylaluminum hydrides in the presence of alkylaluminum alkoxides. It is based on the fact that at moderate temperatures these compounds react with one mole of gaseous ammonia, in accordance with the following equations:

$$AlR_3 + NH_3 \longrightarrow R_2AlNH_2 + RH$$

$$R_2AlH + NH_3 \longrightarrow R_2AlNH_2 + H_2$$

If the resulting dialkylaluminum amides are decomposed with a benzene-alcohol mixture, ammonia is evolved in an amount equivalent to one valency bond of aluminum:

$$R_2AlNH_2 + 3C_2H_5OH \longrightarrow Al(OC_2H_5)_3 + 2RH + NH_3$$

The ammonia which is evolved is determined in the usual way. The method is simple and no special apparatus is needed. It is, however, inaccurate, since the distillation of ammonia is accompanied by a partial loss of dialkylaluminum amide, which as a rule affects the reproducibility of the determinations. Moreover, the method is only suitable for the determination of trialkylaluminum compounds and dialkylaluminum hydride in the presence of dialkylaluminum alkoxides. If dialkylaluminum halides are present as well, the method is unsuitable, since they also bind ammonia with formation of stable addition compounds. In such a case AlR_3 and R_2AlCl can be determined on two separate samples. The halogen content is determined first on one of the samples, the dialkylaluminum halide in the second sample is first converted to the alkoxides by adding an equivalent amount of an alcoholate of an alkali metal, after which the content of active aluminum is determined /85/.

Physicochemical methods of determination of alkylaluminum compounds

Trialkylaluminum compounds and dialkylaluminum hydrides are electron-deficient; they accordingly tend to attract donors with a lone electron pair and to complete the octet by the formation of donor-acceptor complexes.

If organoaluminum compounds are oxidized, this tendency is lost, since the oxygen itself donates an electron pair with formation of associated complexes:

$$\begin{array}{c} R_2AlOR \\ \uparrow \\ ROAlR_2 \end{array}$$

Dialkylaluminum halides also tend to form complexes, but these are less stable than the complexes of trialkylaluminum compounds.

Complex formation is the principle of most methods for the determination of the activity of organoaluminum compounds.

Electrical conductivity

Solutions of alkylaluminum compounds in hydrocarbons are very poor conductors (conductivity of the order of 10^{-9} $ohm^{-1} \cdot cm^{-1}$). When complex compounds are formed, the electrical conductivity increases by one or two orders of magnitude. If the electron-donating agent (amines, esters, etc.) is added continuously, a sharp maximum of electrical conductivity will appear at the equivalence point. As a rule, organoaluminum compounds form complexes with amines at the molar ratio of 1:1. Dialkylaluminum hydrides, which form a nonconducting complex with isoquinoline at this ratio, are an exception. If the ratio dialkylaluminum hydride : isoquinoline is 1:2, the resulting complex has a high conductivity. During the titration of diethylaluminum hydride the electrical conductivity remains unchanged until 1 mole of isoquinoline per 1 mole of diethylaluminum hydride has been added; thereafter, the conductivity increases and begins to fall again as soon as 2 moles of isoquinoline have been added. The electrometric titration curve of mixtures of triethylaluminum with diethylaluminum hydride has at least two maxima.

The sharp change in the electrical conductivity of solutions as a result of formation of donor-acceptor complexes is the principle underlying the different varieties of electrometric titration, proposed by several workers /59, 86 — 90/.

Potentiometric titration

Bonitz /59/, who was the first to describe the direct isoquinoline titration, used the capacity of organoaluminum compounds to form complex compounds with some heterocyclic amines. He then improved the design of the apparatus employed in the determination, as a result of which the method has become a routine automatic method, which is mainly employed in routine control of the purity of triethylaluminum /86/. Other workers /87, 88/ also made use of potentiometric titration by isoquinoline. A platinum and a silver electrode were used. In the determination of diethylaluminum chloride in the presence of ethylaluminum dichloride, the former compound gives a complex with quinoline, the formation of which corresponds to a maximum on the potentiometric curve. Ethylaluminum dichloride does not form complexes with quinoline /88/.

The observed potential jump in the titration of triethylaluminum is 300 — 600 mV, which means that the determination can be performed to within 0.2 — 0.4 abs.%. If triethylaluminum is determined together with diethylaluminum hydride, the hydride can be determined to within ± 7 rel.%. The presence of diethylaluminum ethoxide does not interfere. Triethylaluminum etherate is also titrated against isoquinoline, but the potential jump at the equivalence point is smaller. This fact is utilized to increase the accuracy of the determination of diethylaluminum hydride. If ether is added to the solution to be titrated in an amount which is exactly equivalent to the triethylaluminum present, the potential jump of the diethylaluminum hydride determination will be larger /87/. Graevskii /88/ gave the potentiometric titration curves of mixtures of alkylaluminum compounds. Good results were obtained in potentiometric titrations against organic bases such as pyridine /90/ and pyridine derivatives /89/, especially disubstituted and trisubstituted, since their solutions in benzene remain anhydrous and stable for a long time. Alkylaluminum compounds are determined by 1-, 2- and 3-methylpyridine, 2-ethylpyridine, 2-methyl-5-ethylpyridine, 4-ethylpyridine, or 2, 6, 6-trimethylpyridine. Diethylaluminum hydride and ethylaluminum dichloride cannot be satisfactorily titrated by these reagents. The resulting complexes typically have a molar component ratio of unity. The titration is effected between a platinum and a silver electrode. Owing to the special design of the silver electrode the potential becomes stabilized immediately after the reagents have been added; the potential difference at the equivalence point is 250 — 400 mV. The maximum error is ± 0.1 — 0.2% /89/. This study gives the titration curves against certain ternary bases, at large potential differences.

Conductometric titration

Bonitz /59/ and Graevskii /88/ also studied the conductometric titration of these compounds. As was seen to be the case in the potentiometric titration of triethylaluminum and diethylaluminum hydride against isoquinoline, two conductivity maxima are noted, from which the content of each compound can be found. However, the results thus obtained can be erroneous /83/. Dialkylaluminum halides form conducting complexes with isoquinoline and with quinoline under the same conditions as trialkylaluminum compounds, so that individual determination of these compounds by conductometry is not possible.

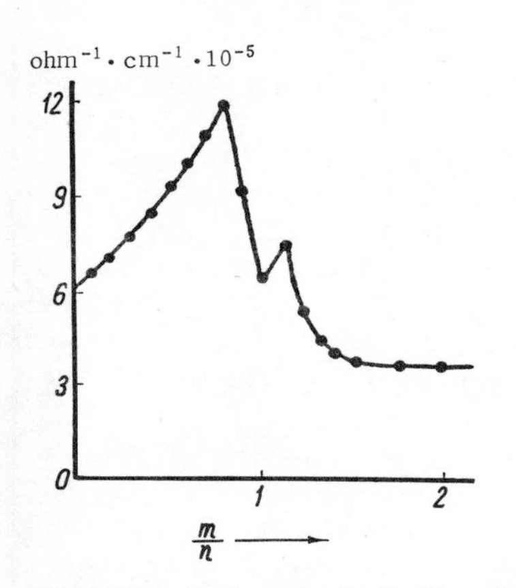

FIGURE 16. Conductometric titration of a mixture of triethylaluminum, diethylaluminum bromide and diethylaluminum ethoxide (m/n – molar ratio of the quinoline consumed in the titration to the aluminum content in the sample).

Graevskii /88/ performed a conductometric determination of diethylaluminum hydride, diethylaluminum bromide and diethylaluminum ethoxide, dissolved in cyclohexane; the titrant was quinoline and the titration was performed in a cell with nonplatinized platinum leaf electrodes, in an inert gas atmosphere. The cell was fed with 1.5 — 30 volt DC. The current intensity was measured with a microammeter or with an M-21 mirror galvanometer. The results obtained by conductometric titration were confirmed by potentiometric titration, but the conductivity maximum was not as sharp as the emf jump during potentiometric titration. Figure 16 shows the curve of conductometric titration of a mixture of triethylaluminum, diethylaluminum hydride, diethylaluminum bromide, and diethylaluminum ethoxide. The rise in the conductivity between the coordinate origin and the first maximum is due to the formation of high-conducting complexes of quinoline with triethylaluminum and diethylaluminum bromide. At the first maximum point all triethylaluminum and diethylaluminum bromide have reacted. The subsequent decrease in conductivity is due to the formation of the nonconducting complex $(C_2H_5)_2AlH \cdot C_9H_7N$, which reacts with excess quinoline to form the complex $(C_2H_5)_2AlH \cdot 2C_9H_7N$, and the electrical conductivity of the system again increases. The location of the second maximum corresponds to the full conversion of the hydride to the latter complex with quinoline. When more quinoline is added, the conductivity of the mixture decreases, since quinoline is a nonconductor, while diethylaluminum ethoxide does not form a complex with quinoline.

Determination of dielectric constant

During donor-acceptor complex formation the donor contributes both electrons to the bond and the electric charge symmetry is impaired. The donor acquires a partial positive charge, while the acceptor acquires a partial negative charge:

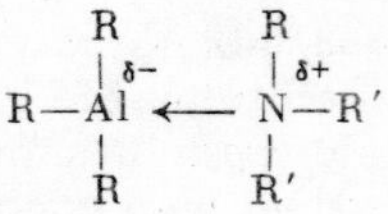

$$\begin{array}{ccc} R & & R \\ |\,\delta- & & |\,\delta+ \\ R\text{—}Al & \longleftarrow & N\text{—}R' \\ | & & | \\ R & & R' \end{array}$$

Owing to the nonsymmetrical charge distribution in the donor-acceptor complexes, the resulting dipole moments are large, $4-6\ D$ /91 — 94/. When brought into contact with oxygen or with water, these complexes again decompose to give compounds with a dipole moment which is very small or nil. The degree of complex formation and the magnitude of the resulting dipole moments depend on the nature both of the donor and of the acceptor. Since the dielectric constant of the solution depends on the nature and on the number of the polar molecules it contains, the determination of the dielectric constant of the solvent will yield the concentration of the molecules formed and the magnitude of the dipole moment /92, 93/.

It has already been mentioned that high dipole moments ($4-6D$) which result from the addition of ethers and amines to compounds of the type AlR_3, R_2AlH, R_2AlHal, etc., may be used to determine the so-called activity by complexometric titration of the dielectric constant. The gradual addition of the donor to the solution of the organoaluminum compound results in a typical peak at the molar ratio of 1:1 and in a typical steep rise of the dielectric constant curve.

In this way it is possible to determine various alkylaluminum compounds in the presence of each other (AlR_3 in the presence of R_2AlH, AlR_3 in the presence of R_2AlHal), since the dipole moments of their addition products differ from one another. It is also possible to make a direct determination of complexes which have not been isolated, if their dipole moment is larger than that of the other components /95/.

Determination of alkylaluminum compounds by titration of the dielectric constant is performed in a special vessel, in which 50 ml of the sample solution are placed and the constants are then thermostated /95/. The titration cell is a Teflon-insulated condenser, with a working capacity of 7 pF, in a circuit with a DC-meter, and can determine dielectric constants between 1 and 7.*

The titrant is added from an automatic buret, the tip of which dips into the solution being titrated. The components are efficiently stirred by a magnetic stirrer. In this way the graphic representation of the results and the calculation of the equivalence values of the titration curves are greatly facilitated. During the determination of the dipole moment, about 3 minutes

* In the initial mixtures containing 20 — 30 mmoles of the alkylaluminum compound in 50 ml of solvent, the variation of the dielectric constant is 50 — 100% of the dielectric constant of the solvent, which is fully sufficient for an accurate determination.

must be left to elapse between the addition of the titrant and the measurement of the dipole moment, in order to allow the heat of reaction to dissipate.

Calorimetric method

The calorimetric method is based on the tendency of trialkylaluminum compounds, dialkylaluminum hydrides and dialkylaluminum halides to react rapidly with electron donor compounds having an active hydrogen atom, with formation of complexes. Most of these reactions are strongly exothermal (10 — 20 kcal/mole). The liberated heat of reaction can be easily recorded if the reaction goes to completion, if the reaction rate is much larger than the titration rate and if the inertia of the titration rate corresponds to the inertia of temperature changes. Hoffmann and Tornow suggested that the calorimetric method be employed in the analysis of organoaluminum compounds. They described a method for the determination of the content of alkylaluminum compounds by calorimetric titration against amines and alcohols, involving the use of an automatic recording apparatus /96/. The determination is based on the formation of molecular compounds with ethers and with tertiary amines and on the solvolysis by alcohols. More than one compound may be determined on the same sample.

The authors also applied their method to the determination of the heat effects of the reaction. The presence of inactive (in this case, oxidized) compounds does not interfere. Everson and Ramirez /97/ developed a method for a simple, rapid and accurate determination of alkylaluminum compounds by calorimetric titration of small samples. In particular, R_3Al, R_2AlH, R_2AlCl, $RAlCl_2$ and R_2AlOR were titrated against triethylamine, isoquinoline, 2, 2'-dipyridyl, dibutyl ether, acetone, tert-butanol, benzophenone, etc. These workers also published the heats of reaction of a number of alkylaluminum compounds (Table 10). Their own results in conjunction with the literature data made it possible to arrive at certain general conclusions as regards the applicability of individual compounds as titrants. It was shown, for example, that one titration of a mixture of a trialkylaluminum compound with dialkylaluminum hydride against amines will determine the contents of AlR_3, of R_2AlH and also their overall activity. However, the determination of alkylaluminum hydrides is not very accurate, especially if the amounts to be determined are small. If the sample contains a large amount of a dialkylaluminum hydride in the presence of a trialkylaluminum compound, titrations against 2, 2'-dimethylpyridine or triethylamine gives satisfactory results for the sum total of the components, but not for the contents of individual compounds. If the titrant employed is isoquinoline, the determination of the dialkylaluminum hydride is even less accurate. Diphenyl ether and anisole react with AlR_3 with evolution of a very small amount of heat and cannot be used as calorimetric titrants. Titrations of individual AlR_3 and R_2AlH compounds against ethers and ketones give good results; the hydrides are determined much more accurately, and the reaction is more sensitive than in the titration against amines. The most suitable titrant is tert-butanol, since it reacts with AlR_3 and with R_2AlH in one distinct step. The higher the molecular weight of the alcohol used as titrant, the slower will be the reaction and the less satisfactory the results obtained by calorimetric titration.

TABLE 10. Heats of reaction of alkylaluminum compounds /56, 96, 97/

Compound	Heat evolved (kcal/mole) in the reaction with			
	dibutyl ether	tert-butanol	triethylamine	isoquinoline
$(CH_3)_3Al$	11	41	14	20
			15.4	
$(CH_3)_2AlH$	—	—	10.1	—
$(C_2H_5)_3Al$	13	51	18	19
	11.0		15.2	20.7
$(C_2H_5)_2AlH$	—	—	11	—
$(C_3H_7)_3Al$	—	—	14.6	—
$(iso\text{-}C_3H_7)_3Al$	17	45	18	23
$(C_4H_9)_3Al$	—	—	14.6	—
$(iso\text{-}C_4H_9)_3Al$	18	44	22.4	23
$(iso\text{-}C_4H_9)_2AlH$	2	37	—	25
$(C_6H_{13})_3Al$	14	—	—	—
$(C_8H_{17})_3Al$	12	38	14.0	13
$(CH_3)_2AlCl$	—	—	15.8	—
$(C_2H_5)_2AlCl$	11	39	16	17
			16.0	17
$C_2H_5AlCl_2$	16	44	24	21

Colorimetric method

Trialkylaluminum compounds react with compounds such as pyridine, isoquinoline, benzalaniline, etc., to give colored complexes, which make it possible to determine small amounts of alkylaluminum compounds. Trialkylaluminum compounds give colored molecular compounds in the ratio of 1:1, as follows /98/:

$$AlR_3 + C_7H_9N \longrightarrow AlR_3 \longleftarrow NC_7H_9$$

$$AlR_3 + RCH{=}NR \longrightarrow AlR_3 \longleftarrow N(R){=}CH{-}R$$

Dialkylaluminum hydrides react to form colorless complexes if the molar ratio is 1:1; colored complexes are given if the molar ratio is 1:2. The color intensity of the complexes depends on the nature of the azomethine bonds:

$$R_2AlH + RCH{=}NR \longrightarrow R{-}CH_2{-}N(AlR_2){-}R$$

$$R{-}CH_2{-}N(AlR_2)R + RCH{=}NR \longrightarrow \begin{array}{c} R{-}CH_2{-}N(AlR_2){-}R \\ \uparrow \\ R{-}CH{-}NR \end{array}$$

The acid properties of organoaluminum compounds make it possible to effect the titration in the presence of ordinary acid-base indicators; these may be used in the determination of the equivalence point in the titration of alkylaluminum compounds against electron-donating reagents such as aromatic amines and ethers /32/. It was noted /99/ that when an alkylaluminum compound is added to the indicator (Methyl Violet), the latter changes color from violet (basic form) to yellow or green (acid form). When the base is again added, the color turns back to violet:

$$[In + AlR_3] \longrightarrow [AlR_3—In]$$

$$[AlR_3—In] + C_6H_5—N: \longrightarrow [AlR_3 \longleftarrow :N—C_6H_5] + In$$

Dialkylaluminum alkoxides do not form stable complexes with these reagents and do not change the color of the indicator. Bonitz /59/ utilized the red-colored complex of isoquinoline : dialkylaluminum hydride = 2 : 1 in the capacity of indicator in the volumetric titration of active trialkylaluminum compounds against isoquinoline. This method is suitable only if the added dialkylaluminum hydride behaves exclusively as indicator.

Razuvaev and Graevskii /99/ developed an indicator method based on the titration of alkylaluminum compounds against dimethylaniline, pyridine, butyl acetate, ethyl acetate and ethyl ether in the presence of indicators which are readily soluble in organic solvents and give sharp equivalence points (e.g., Methyl Violet, Crystal Violet, Gentian Violet). The titration was conducted in the presence of excess of the organic base, which was determined by the spectrophotometric method, using the IM-2 monochromator. If this is done, all indicators used in the volumetric determination of the concentration of alkylaluminum compounds act as visual indicators.

Other workers, who use triphenylmethane type indicators in the titration of organoaluminum compounds by electron-donating reagents of a similar type, found that most alkylaluminum compounds could be accurately and rapidly determined by this technique /100/. Triphenylmethane indicators can act reversibly if the amino groups are not fully methylated.

Titration in the presence of visual indicators is applicable to dilute solutions of alkylaluminum compounds which are practically free from dialkylaluminum hydrides; this is because concentrated (above 30%) solutions of alkylaluminum compounds decompose the indicators in the presence of a dialkylaluminum hydride /99, 100/.

Colored complexes of alkylaluminum compounds with nitrogenous compounds (e.g., benzalaniline, pyridine, isoquinoline, etc.) have characteristic absorption spectra. This fact may be utilized in spectrographic determination of a number of organoaluminum compounds /61, 98, 100 — 102/.

Neumann /61/ developed a spectrophotometric method for the determination of dialkylaluminum hydrides, in which the alkyl radical ranged between ethyl and dodecyl. The experimental error is ± 2 rel.% and the sensitivity is 0.1%, if isoquinoline or benzalaniline are employed as the reagents. This method was used for several years on laboratory scale in various kinds of investigations and has also proved itself in industrial practice.

Mitchen /101/, who studied the different stabilities of the colored complex bonds $R_2AlH \leftarrow$ isoquinoline and $AlR_3 \leftarrow$ isoquinoline, proposed a spectrophotometric method for a complete analysis of trialkylaluminum compounds and dialkylaluminum hydrides. This method is applicable to trialkylaluminum compounds only if the R_2AlH has been added to the sample as indicator. The author pointed out that his method is relatively simple and accurate (± 3 rel. %) and found that the results agreed with those obtained by hydrolysis followed by gas analysis. The method was improved somewhat by Hagen and Leslie /100/.

Wadelin /102/ reported a spectrophotometric isoquinoline titration method and applied it to the analysis of AlR_3, in which the alkyl group ranged from methyl to hexyl. The spectrophotometric measurements were carried out at 460 mμ. This wavelength is utilized to determine dialkylaluminum hydrides which contain trialkylaluminum compounds.

II. ANALYSIS OF SALTS OF TRANSITION METALS

All salts of transition metals employed as constituents of catalyst complexes are readily soluble in hydrochloric and sulfuric acids. Therefore, these compounds are most often analyzed by determining the metal ion in solution by any known method used in ore analysis /103 — 106/.

The most frequently employed compounds of this type are titanium salts. It is often important to know not merely the total titanium content, but also the contents of titanium ions of the different valencies. Accordingly, the determination of titanium in the various valency states, and the main methods used in the analysis of other transition elements will be the main subjects of the sections which follow.

1. DETERMINATION OF TITANIUM IN THE VARIOUS VALENCY STATES

Titanium may be determined by volumetric, photocolorimetric and polarographic methods of analysis.

The volumetric method is a simple technique for the determination of total titanium. It is based on the reduction of tetravalent titanium in Jones' reductor to trivalent titanium; this is followed by titration of trivalent titanium against ferric ammonium alum in the presence of ammonium thiocyanate as indicator. After all the trivalent titanium has been oxidized by the ferric salt, the first drop of the alum in excess forms the red $Fe(CNS)_3$, which indicates the end of the titration /107/. In order to prevent any oxidation of trivalent titanium by atmospheric oxygen, it is recommended that a small amount of saturated ammonium sulfate solution or glacial acetic acid be added to the solution. These reagents form stable complexes with Ti(III), as a result of which there is a large positive potential shift of the system Ti^{4+}/Ti^{3+} /108/. Alternatively, the titration of trivalent titanium may be effected in an atmosphere of carbon dioxide or nitrogen /105/.

Trivalent and tetravalent titanium may be determined in the presence of each other. Trivalent titanium is determined directly; this is followed by the determination of the total titanium after reduction of Ti(IV) to Ti(III) in Jones' reductor. Trivalent titanium is determined by the method proposed by Tabakova and Solov'eva /109/, by introducing a known volume of a standard solution titanium salt into an accurately known amount of a solution of ferric ammonium alum in excess, after which the ferrous iron formed is titrated against permanganate. The content of the various forms of titanium is calculated from the consumption of the permanganate in the first and second titrations.

The results of determinations of titanium of various valencies by this method showed satisfactory agreement with the results of cerimetric determination of trivalent titanium /110/ and those of colorimetric determination of tetravalent titanium /109/.

The photocolorimetric method for the determination of titanium is based on the formation of a yellow complex of titanium with hydrogen peroxide in an acid medium /104/. The color intensity of the complex is independent of the acidity of the medium. The photometric determination is made using a blue filter.

In another method, Ti(III) and Ti(IV) are determined in aqueous ether solutions by taking advantage of the reaction between titanium ions and potassium thiocyanate /111, 112/. In neutral medium trivalent titanium ions are the only ones to form a dark-violet complex with potassium thiocyanate. In acid medium both Ti(III) and Ti(IV) form a dark-red complex.

It was shown by a number of workers that titanium of different valencies can be determined by polarography. Zeltzer /113/ was the first to obtain a sharp reduction wave of Ti(IV) in dilute solutions of hydrochloric, sulfuric and nitric acids. According to Strubl /114/, polarographic determination of Ti(IV) in acid medium is not to be recommended, since titanium salts are hydrolyzed in weakly acid solutions, whereas in strongly acid solutions the wave is unsharp and merges with the wave corresponding to the discharge of hydrogen ions. This was also confirmed by Sinyakova /115/. It was found, on the other hand, that at a certain acidity value (0.1 — 0.5 N solution) the height of the titanium wave is exactly proportional to its concentration in solution /116/. It was subsequently shown by Krylov et al. /117/ that Ti(IV) may be directly determined by polarography in sulfuric acid solution. Polarographic determination of Ti(IV) can also be effected in the presence of complexants, when sharp waves, corresponding to the reversible reaction $Ti^{4+} \rightleftharpoons Ti^{3+}$, are obtained. The reduction potential of titanium becomes more positive. Trilon B /115/, thiocyanates /118/ and oxalates /119, 120/ can be employed as complexants in the determination of titanium.

Khoroshin and Fikhtengol'ts /121/ carried out the most detailed study of the polarographic determination of bivalent, trivalent and tetravalent titanium. They showed that titanium ions of different valencies can be determined by polarographic determination of Ti(III) and Ti(II) in two different samples, prepared under different conditions. One of the samples is dissolved in excess ferric ammonium alum in an inert gas atmosphere. Under these conditions Ti^{2+} and Ti^{3+} are oxidized to Ti^{4+}:

$$Ti^{2+} + 2Fe^{3+} \longrightarrow Ti^{4+} + 2Fe^{2+}$$

$$Ti^{3+} + Fe^{3+} \longrightarrow Ti^{4+} + Fe^{2+}$$

The amount of the ferrous ion formed, which is equivalent to the sum of Ti^{2+} and Ti^{3+}, is determined by the polarographic technique. A second sample of the catalyst is dissolved in 5 N HCl in a continuous stream of an inert gas; the acid solution contains ammonium sulfate as the stabilizer of trivalent titanium. Under these conditions Ti^{2+} is converted to Ti^{3+} in accordance with the equation:

$$TiCl_2 + HCl \longrightarrow TiCl_3 + {}^1/_2H_2$$

The polarographic analysis of the solution thus yields the content of Ti^{4+} in the catalyst, as well as the sum of bivalent and trivalent titanium in the form of Ti^{3+}. If two samples, containing a mixture of titanium ions of various valencies, are analyzed, the content of Ti^{2+} and Ti^{3+} in the catalyst can be determined. If Ti^{2+} is absent, the analysis becomes much simpler, and Ti^{3+} and Ti^{4+} can be determined on the same sample.

According to the authors, the average error involved in the determination of Ti^{3+} and Ti^{4+} is about 3.0%. The duration of the entire determination does not exceed 2 — 3 hours. Aluminum present in the sample does not interfere.

2. PRINCIPAL METHODS OF ANALYSIS OF SALTS OF OTHER TRANSITION METALS

Zirconium ions may be determined gravimetrically (e.g., by the phosphate or by the phenylarsenate method), as described in detail in the case of silicate rocks /103/. Zirconium can also be determined by the photocolorimetric method; the most sensitive reagent for this purpose is Xylenol Orange. It forms a yellow-red complex with zirconium, which is stable in moderately strong acid medium /104/. The color intensity is determined using an FEK-N-57 photocolorimeter with a No. 5 filter. The analytical error is ± 0.002% Zr. Ferric ion and hexavalent molybdenum interfere, since they, too, form colored compounds with Xylenol Orange.

Vanadium may be determined by colorimetric, gravimetric or volumetric methods, depending on the concentration and on the presence or otherwise of other elements in the sample. The colorimetric method is most often employed /122, 123/. The volumetric method is used for vanadium contents of the order of one percent or less /105/.

The colorimetric method is based on the formation of the greenish-yellow phosphorus-tungsten-vanadium complex when phosphoric acid and sodium tungstate are added to an acid solution which contains vanadium. The color of the complex at once reaches the maximum intensity, if the concentration of sodium tungstate in solution is not less than 0.003 mole/liter. The order in which the reagents are added is important: phosphoric acid is added first, and is followed by sodium tungstate. The optimum concentration of phosphoric acid is 0.1 M, while that of sodium tungstate is 0.006 M. Colorimetric determinations are made in an FEK-M photocolorimeter with a blue filter (SS-8). Pentavalent vanadium may also be photometrically determined by aminophenol /123a/. The method is applicable to vanadium contents ranging from 10^{-3}% to 2 — 3%. Another original colorimetric method for the determination of tetravalent vanadium, developed by Ponomarev and Ratina /123/, is based on the redox reaction between tetravalent vanadium and ferric ion. This reaction takes place in alkaline and weakly alkaline media:

$$V^{4+} + Fe^{3+} \longrightarrow V^{5+} + Fe^{2+}$$

The ferrous iron formed during the reaction can be very accurately determined colorimetrically using α, α'-dipyridyl, and the content of vanadium thus calculated.

Ferrous iron reacts with α, α'-dipyridyl to form the soluble, intensely red-colored complex $[Fe(C_{10}H_8N_2)_3]^{2+}$. The reaction is very sensitive and specific. The color is stable to atmospheric oxygen and attains the maximum intensity at pH 2.5 — 9. In more acid solutions the color develops slowly and does not attain the maximum intensity.

The most popular volumetric method for the determination of vanadium is that proposed by Syrokomskii and Stepin /105/. The method is based on the redox reaction which takes place between the vanadium compound in sulfuric acid (at least 5N) and Mohr's salt:

$$(VO)_2(SO_4)_3 + 2FeSO_4 \longrightarrow 2VOSO_4 + Fe_2(SO_4)_3$$

Phenylanthranilic acid is used as indicator. The solution of the vanadium compound becomes intense pink-violet. When the solution is titrated against 0.02 M solution of Mohr's salt $(NH_4)_2SO_4 \cdot FeSO_4 \cdot 6H_2O$, the pink solution turns greenish-yellow on the addition of the last drop. The disadvantage of the method is that the titer of the solution of Mohr's salt must be verified every day.

The determination of hexavalent chromium is based on the oxidation of diphenylcarbazide by the metal in acid solution. The resulting compound colors the solution red-violet. The intensity of the resulting coloration is matched against that of a standard solution of chromium. If an insufficient amount of the reagent is added, the hexavalent chromium continues to oxidize the colored compound and the color is discharged /104/.

The most frequently employed volumetric method involves a preliminary oxidation of chromium to Cr(VI) (usually by potassium persulfate or sodium peroxide), which is then reduced to Cr(III) by the addition of a standard solution of ferrous sulfate. The excess ferrous sulfate is titrated against potassium permanganate solution until the green solution turns pale pink. The hexavalent chromium may also be titrated directly by ferrous sulfate in the presence of phenylanthranilic acid as indicator:

$$K_2Cr_2O_7 + 6FeSO_4 + 7H_2SO_4 \longrightarrow Cr_2(SO_4)_3 + 3Fe_2(SO_4)_3 + K_2SO_4 + 7H_2O$$

Both these methods yield equally accurate results /108/.

The best method for the determination of manganese is the potentiometric titration, in which Mn(II) is oxidized to Mn(III) by potassium permanganate in pyrophosphate solution /124/. The method may be used to determine both large and small amounts of manganese. If the content of manganese is smaller than 1%, it is oxidized to permanganic acid by alkali periodate or ammonium persulfate in nitric acid, sulfuric acid or, better, phospho-sulfuric acid medium, and the color is matched against that given by a standard manganese solution /104/. Up to 0.001% manganese can be determined in this way.

If the content of manganese in solution exceeds 1%, it may be determined by a volumetric method. These include the persulfate, persulfate-arsenite and other methods. The persulfate method, which is the one most frequently employed, involves the oxidation of Mn(II) to Mn(VII) by ammonium persulfate in the presence of silver nitrate as catalysts:

$$2MnSO_4 + 5(NH_4)_2S_2O_8 + 8H_2O \longrightarrow 2HMnO_4 + 5(NH_4)_2SO_4 + 7H_2SO_4$$

The oxidation is carried out in sulfuric or nitric acid solution in the presence of phosphoric acid as stabilizer /104/.

The large number of publications on the determination of cobalt notwithstanding, many of the methods which have been proposed are unsatisfactory. The colorimetric and polarographic techniques are the most rapid. Thus, the colorimetric pyrophosphate-thiocyanate method will readily and rapidly determine cobalt concentrations between 0.01 and 1%. The method is based on the formation of a cobalt-thiocyanate complex in neutral or weakly acid medium. The complex is soluble in organic solvents (acetone, amyl alcohol, ether). The undissociated complex colors the solution an intense blue. The optimum conditions for the formation of the complex and an accurate determination of cobalt are ensured by introducing a small amount of sodium pyrophosphate /106/. Cobalt may be colorimetrically determined by nitroso-R-salt (1-nitroso-2-naphthol-3, 6-disulfonic acid, sodium salt), which oxidizes Co(II) to Co(III) and colors the solution red owing to complex formation /125, 126/. Cobalt is polarographically determined with ammonium chloride as supporting electrolyte. Trivalent cobalt gives two waves, corresponding, respectively, to the reductions $Co^{3+} \rightarrow Co^{2+}$, and $Co^{2+} \rightarrow Co^{0}$. The second wave $E_{1/2} \cong -1.3$ volt, is used in the determination. Cobaltous cobalt gives only one wave, with $E_{1/2} = -1.3$ volt /127/.

Nickel is most often determined by photometric and trilonometric methods. In the photometric method, the color intensity of the brown-red complex of nickel with dimethylglyoxime /128/ is measured in alkaline medium in the presence of an oxidizing agent (ammonium persulfate, bromine water, etc.). The color intensity may be determined visually or photometrically. The method is applicable if the nickel content is 0.0005 — 3%.

The trilonometric determination of nickel involves an ordinary titration of a hydrochloric acid solution of nickel against a solution of Trilon B, with murexide as indicator. Nickel reacts with murexide to form an intense yellow-colored complex. During the titration against Trilon B, this complex is decomposed; nickel is complexed by Trilon B with displacement of the murexide and the solution becomes pink-violet at the equivalence point /104/.

Bibliography

1. Gel'man, N.E. and I.I. Bryushkova. — ZhAKh, 19(3):369. 1964.
2. Bähr, G. and H. Müller. — Chem. Ber., **88**:251. 1955.
3. Kreshkov, A.P., V.A. Bork, E.A. Bondarevskaya, et al. Prakticheskoe rukovodstvo po analizu monomernykh i polimernykh kremnii-organicheskikh soedinenii (Practical Guide to the Analysis of Monomeric and Polymeric Organosilicon Compounds). — Goskhimizdat. 1962.
4. Franklin, A. and S. Voltz. — Analyt. Chem., **27**:865. 1955.
5. Head, E. and Ch. E. Holley. — Analyt. Chem., **28**:1172. 1956.
6. Pickhardt, W., L. Safranski, and J. Mitchell. — Analyt. Chem., **30**:1298. 1958.
7. Mitsui, P. — J. Soc. Org. Synt. Chem. Japan, **17**:474. 1959.
8. Williams, A. and T.O. Park. — Analyst, **85**:126. 1960.
9. Korshun, M.O. and E.A. Bondarevskaya. — In: "Trudy nauchno-issledovatel'skogo instituta," p. 127. Goskhimizdat. 1958.
10. Korshun, M.O. and N.E. Gel'man. Novye metody elementarnogo mikroanaliza (New Methods of Element Microanalysis). — Goskhimizdat. 1949.
11. Korshun, M.O., N.E. Gel'man, and N.S. Sheveleva. — ZhAKh, **13**:695. 1958.
12. Gel'man, N.E. and N.S. Sheveleva. — ZhAKh, **20**(6):719. 1965.
13. Gel'man, N.E., M.O. Korshun, et al. — Doklady AN SSSR, 129:1046. 1959.
14. Gel'man, N.E. and Wang Wên-yün. — ZhAKh, **15**:487. 1960.
15. Korshun, M.O. and V.A. Klimova. — ZhAKh, **3**:274. 1947.
16. Korshun, M.O. — ZhAKh, **7**:96. 1952.
17. Klimova, V.A. and T.A. Antipova. — ZhAKh, **16**:465. 1961.
18. Ingram, G. — Analyst, **86**:411. 1961.

19. Gel'man, N.E. et al. — Doklady AN SSSR, **161**:107. 1965.
20. Korshun, M.O., N.S.Sheveleva, and N.E.Gel'man. — ZhAKh, **15**:99. 1960.
21. Grosse, A. and J.Mavity. — J.Org.Chem., **5**:106. 1940.
22. Zhigach, A.F. et al. — Khimicheskaya Promyshlennost', **2**:123. 1959.
23. Kryukov, S.I., M.A.Kutin, and M.I.Farberov. — Izvestiya Vuzov. Khimiya i khimicheskaya tekhnologiya, **1**:86. 1958.
24. Dahling, V. and S.Pasynkiewicz. — Przemysl Chemiczny, **39**:300. 1960.
25. Welcher, I. The Analytical Uses of Ethylenediamino Tetraacetic Acid, Vol.I. 1958.
26. Wänninen, E. and A. Ringbom. — Anal. Chim. Acta, **12**:308. 1955.
27. Schwarzenbach, H. Die Komplexmetrische Titration, Die Chemische Analyse. — Stuttgart, pp. 45 — 85. 1960.
28. Přibil, R. Komplexony v chemické analyse. — Praha, Publ. House of the Czechoslovak Acad. of Sciences. 1953.
29. Welcher, I. Organic Analytical Reagents, Ed. 3, Vol.I, Toronto—New York — London, p. 283. 1955.
30. Chernokhov, A.Yu., N.N.Dobkina, and M.L.Khersonskaya. — Zavodskaya Laboratoriya, **21**:638. 1955.
31. Miller, F. et al. Mikrochemie der Mikrochim. acta, **40**(4):373. 1953.
32. Ziegler, K., H.Gellert, et al. — Ann., **589**:91. 1954.
33. Nesmeyanov, A.N. and N.N.Novikov. Sinteticheskie metody v oblasti metallorganicheskikh soedinenii (Synthetic Methods in the Field of Organometallic Compounds), No. 4:89. — Izd. AN SSSR. 1945.
34. Boldyreva, I.I., B.A.Dolgoplosk, and V.A.Krol'. — VMS, **1**(6):900. 1959.
35. Burke, K. and C.Davis. — Analyt. Chem., **36**(1):172. 1964.
36. Babenyshev, V.M. and O.M.Kuznetsova. — ZhAKh, **15**(5):568. 1960.
37. Mikhailenko, M.I. — Sbornik issledovanii v oblasti farmakologii (Collection of Research Papers on Pharmacology), pp. 105 — 112. Odessa. 1959.
38. Wetlesen, C. and S.Omang. — Analyt. Chim. Acta, **24**(3):294. 1961; RZhKhim, 16D69. 1961.
39. Bashkirtseva, A.A. and E.M.Yakimets. — Zavodskaya Laboratoriya, **25**(10):1166. 1959.
40. Babko, A.K. and T.N.Nazarchuk. — Ukrainskii Khimicheskii Zhurnal, **20**(6):678. 1954.
41. Bashkirtseva, L.A. and E.M.Yakimets. — Trudy Ural'skogo Politekhnicheskogo Instituta im. S. M. Kirova, No. 58:76. 1957.
42. Theis, M. — Z. anal. Chem., **144**(2):106. 1955.
43. Bashkirtseva, L.A. Author's Summary of Thesis. Ural'skii Politekhnicheskii Institut. 1956.
44. Flaschka, H., K.Haar, and J.Bazen. — Mikrochim. acta, **4**:345. 1953.
45. Mulner, G. and J.Woodhead. — Analyst, **79**(939):363. 1954.
46. Sajó István. — Acta Chim. Acad. Sci. Hungar., **6**:251. 1955.
47. Flaschka, H. and Z.Franschitz. — Z. anal. Chem., **144**(6):421. 1955.

48. Schwarzenbach, H. — Analyst, **80**(955):713. 1955.
49. Přibil, R., J. Činalik, J. Doležal, V. Simon, and J. Zyka. — Čs. farmak., **2**(7/8):223. 1953.
50. Flaschka, H. and M. Abdine. — Mikrochim. Acta, **1**:37. 1955.
51. Bashkirtseva, L.A. and E.M. Yakimets. — Zavodskaya Laboratoriya, **25**(5):540. 1959.
52. Solomon, O., E. Mihăilescu, and P. Glineschi. — ZhPKh, **36**(12):2712. 1963.
53. Jurczyk, J. — Chem. Analytyczna, **10**(3):441. 1965.
54. Bartkiewicz, S. and J. Robinson. — Anal. Chim. Acta, **20**(4):326. 1959.
55. Crompton, T. — Analyst, **91**:374. 1966.
56. Terent'ev, A.P., G.G. Larikova, et al. — ZhAKh, **18**:514. 1963.
57. Crompton, T. and V. Reid. — Analyst, **88**:713. 1963.
58. Dijkstra, R. and E. Dahmen. — Z. anal. Chem., **181**:399. 1961.
59. Bonitz, E. — Chem. Ber., **88**:742. 1955.
60. Bernard, H. — XVI Congres int. Chimie pure et appl., **2**:122. 1957.
61. Neumann, W. — Lieb. Ann. Chem., **629**:23. 1960.
62. Lionozova, R.Z. and M.L. Genusov. — Zavodskaya Laboratoriya, **26**(8):945. 1960.
63. Shvindlerman, G.S. and E.N. Zavadovskaya. — Zavodskaya Laboratoriya, **31**(1):32. 1965.
64. Bondarevskaya, E.A., S.V. Syartsillo, and R.N. Potsepkina. — Trudy komissii po analiticheskoi khimii. Organicheskii Analiz, **13**:178. Izd. AN SSSR. 1963.
65. Crompton, T. — Analyst, **86**(1027):652. 1961.
66. Reid, V. and R. Truelove. — Analyst, **77**:325. 1952.
67. Reid, V. and D. Salmon. — Analyst, **80**(954):704. 1955.
68. Kratochvil, V. and C. Soběslavský. — Chem. prumysl, **6**(12): 515. 1956; No. 16:54751. 1957.
69. Griffiths, V. and D. Stock. — J. Chem. Soc., p. 1633. 1956.
70. Bondarevskaya, E.A., S.V. Syartsillo, and R.N. Potsepkina. — ZhAKh, **14**(4):501. 1948.
71. Pudovik, A.N. and G.M. Sinaiskii. — ZhPKh, **21**(8):862. 1948.
72. Nakhmanovich, B.M. and N.V. Pryanishnikova. — Zavodskaya Laboratoriya, **23**(2):165. 1957.
73. Yüeh Kuo-ts'ui. — Galfan' Tszy Tung hsün, **2**(3):192. 1958.
74. Badin, E. — J. Phys. Chem., **63**:1791. 1959.
75. Tepenitsyna, E.P. and M.I. Farberov. — Izvestiya Vuzov. Khimiya i khimicheskaya tekhnologiya, **4**:765. 1958; Tepenitsyna, E.P., M.I. Farberov, A.M. Kut'in, and G.S. Levskaya. — Vysokomolekulyarnye Soedineniya, **8**:1148. 1959.
76. Murahashi, S. — Bull. Chem. Soc. Japan, p. 1094. 1959.
77. Arlman, E. and J. de Jong. — Rec. trav. chim., **8**:910. 1960.
78. Habvinga, R. and J. de Jong. — Rec. trav. chim., **1**:56. 1960.
79. Orzechowsky, A. — J. Polymer Sci., **34**:65. 1959.
80. Simon, A. and L. Kovacs. — Magyar Kém. lapja, **13**:180. 1958.
81. Herman, D. — J. Am. Chem. Soc., **75**:3883. 1953.
82. Kryukov, S.I., A.M. Kut'in, G.S. Levskaya, E.P. Tepenitsyna Z.F. Ustavshchikov, and I.I. Farberov. — Izvestiya Vuzov. Khimiya i khimicheskaya tekhnologiya, Vol. 86. 1958.

83. Crompton, T. — Analyt. Chem., **39** (2):268. 1967.
84. Tabakova, E.G. and Z.V. Solov'eva. — Zavodskaya Laboratoriya, **22**:1417. 1956.
85. Ziegler, K. and H. Gellert. — Lieb. Ann. Chem., **629**:20. 1960.
86. Bonitz, E. and W. Huber. — Z. anal. Chem., **186**:206. 1962.
87. Farina, M. et al. — Ann. Chim., **48**:501. 1958.
88. Graevskii, A.S. — Doklady AN SSSR, **119** (1):101. 1958.
89. Nebbia, L. and B. Pagami. — Chim. e ind., **44**(5):383. 1962.
90. Uhniat, M. and T. Zawada. — Chem. Anal. (Warsaw), **9**(4):701. 1964.
91. Hoffmann, E. — Z. Elektrochem., Ber. Bunsenges. physik. Chem., **61**:1014. 1957.
92. Hoffmann, E. and G. Schomburg. — Z. Elektrochem., Ber. Bunsenges. physik. Chem., **61**:1101. 1957.
93. Zeiss, H. (editor). Organometallic Chemistry. — New York, Reinhold. 1960.
94. Strohmeier, W. et al. — Z. Elektrochem., Ber. Bunsenges. physik. Chem., **61**:1010. 1957.
95. Hoffmann, E. and W. Tornow. — Z. analyt. Chem., **186**:231. 1962.
96. Hoffmann, E. and W. Tornow. — Z. analyt. Chem., **188**:321. 1962.
97. Everson, W. and E. Ramirez. — Analyt. Chem., **37**(7):806. 1965.
98. Neumann, W. — Lieb. Ann. Chem., **618**:90. 1958.
99. Razuvaev, G.A. and A.S. Graevskii. — Doklady AN SSSR, **128**:309. 1959.
100. Hagen, D. and W. Leslie. — Anal. Chem., **35**(7):814. 1963.
101. Mitchen, J. — Anal. Chem., **33**(10):1331. 1961.
102. Wadelin, C. — Talanta (London), **10**:97. 1963.
103. Ponomarev, A.I. Metody khimicheskogo analiza silikatnykh i karbonatnykh gornykh porod (Methods of Chemical Analysis of Silicate and Calcareous Rocks). — Izd. AN SSSR. 1961.
104. Ponomarev, A.I. Metody khimicheskogo analiza zheleznykh, titanomagnievykh i khromovykh rud (Methods of Chemical Analysis of Iron, Titano-Magnesium, and Chrome Ores). — Izd. "Nauka." 1966.
105. Syrokomskii, V.S. Metody analiza zheleznykh i margantsevykh rud (Methods of Analysis of Iron and Maganese Ores). — Metallurgizdat. 1950.
106. Zvenigorskaya, V.M. Metody opredeleniya kobal'ta i margantsa (Determination of Cobalt and Manganese). — Gosgeoltekhizdat. 1946.
107. Kolthoff, I.M. et al. Volumetric Analysis, Vol. 3. — New York, Interscience. 1942—57.
108. Syrokomskii, V.S., E.V. Silaeva, and V.B. Avilov. — Zavodskaya Laboratoriya, **15**:896. 1949.
109. Tabakova, E.G. and Z.V. Solov'eva. — Zavodskaya Laboratoriya, **12**:1417. 1956.
110. Syrokomskii, V.A. and E.V. Silaeva. — Zavodskaya Laboratoriya, **15**(9):1015. 1949.
111. Patshene, G. and W. Schnaller. — Z. anorg. Chem., **235**:257. 1938.
112. Fresenius, W. and G. Jander. — Handbuch der analytischen Chemie, Part IV, Vol. IVb, Vc:221. 1956.
113. Zeltzer, S. — Coll. Czechosl. Chem. Comm., **10**:475. 1938.

114. Strubl, R. — Coll. Czechosl. Chem. Comm., **10**:475. 1938.
115. Sinyakova, S.I. — ZhAKh, **8**(6):333. 1953.
116. Zan'ko, A.M. et al. — Zavodskaya Laboratoriya, **9**(9):976. 1940.
117. Krylov, E.I. et al. — Doklady AN SSSR, **98**:593. 1954.
118. Kolousek, M. — Coll. Czechosl. Chem. Comm., **11**:592. 1939.
119. Adams, D. — Anal. Chem., **20**:891. 1948.
120. Vandenbosch, V. — Bull. Soc. Chim. Belges, **58**:532. 1949; C.A., **44**:9863g. 1950.
121. Khoroshin, A.V. and V.S. Fikhtengol'ts. Fiziko-khimicheskie metody analiza i issledovaniya produktov proizvodstva sinteticheskogo kauchuka (Physicochemical Methods of Analysis and Investigation of the Synthetic Rubber Products), p. 3. — Goskhimizdat. 1961.
122. Vinogradov, A.P. — Doklady AN SSSR, Seriya, **10**:249. 1931.
123. Ponomarev, A.I. et al. — Zavodskaya Laboratoriya, **21**:918. 1955.
124. Slavatinskii, A.S. — ZhAKh, **12**:485. 1957.
125. Malyuga, D.P. — ZhAKh, **1**:176. 1946.
126. Malyuga, D.P. — ZhAKh, **2**:323. 1947.
127. Kolthoff, I.M. and J.J. Lingane. Polarography. — New York, Interscience. 1946.
128. Sbornik "Metody khimicheskogo analiza mineral'nogo syr'ya," VIMS, No. 6, Gosgeoltekhizdat. 1960.

Chapter V

UTILIZATION OF COMPLEX ORGANOMETALLIC CATALYSTS IN POLYMERIZATION

It has been stated in the introduction to this book that complex organometallic catalysts play an important part in polymerization processes. These catalysts are employed in industrial production of high-density polyethylene, polypropylene, and stereoregular rubbers of a quality comparable to that of natural rubber. The scope of application of complex catalysts keeps increasing, as does the number of monomers capable of being polymerized by these catalysts: these include not only hydrocarbons (olefins and dienes), but also a number of polar monomers — vinyl ethers, vinyl chlorides, olefin oxides, alkylene sulfides, etc.

Up till recently /1 — 9/ most of the experimental data and most of the scientific studies published in the literature concerned the complexes of alkylaluminum compounds with vanadium or titanium halides. This group of complexes, more than any other so far studied in the laboratory, has found practical application in the polymerization of α-olefins and several other monomers in industry. The probable reason for it is that the electron structure of the atoms of these transition metals is most favorable to the formation of complexes which are stable and are effective catalysts of stereospecific polymerization processes.

1. TYPES OF ORGANOALUMINUM CONSTITUENTS OF COMPLEX POLYMERIZATION CATALYSTS

The systems alkylaluminum compound — titanium chloride so far studied can be subdivided into four groups, depending on the nature of the constituent organoaluminum compound /10/.

The first group includes compounds of the general formula $Al(C_2H_5)_2X$, where X = C_2H_5, Cl, Br, I, F, OC_6H_5, SC_6H_5, SeC_6H_5, NC_5H_{10}.

The most frequently encountered compounds in this group are $Al(C_2H_5)_3$ and $Al(C_2H_5)_2Cl$, which have been studied in fair detail, especially in connection with the preparation of polypropylene /11, 12/. Organoaluminum compounds such as $Al(C_2H_5)_2SC_6H_5$ and $Al(C_2H_5)_2SeC_6H_5$ react with $TiCl_3$ to form catalysts which are highly stereospecific, almost as much as catalysts based on $Al(C_2H_5)_2Cl$.* On the other hand, compounds such as $Al(C_2H_5)_2OC_6H_5$ and $Al(C_2H_5)_2NC_5H_{10}$ do not form stereospecific catalysts, such as would

* Here and in what follows the measure of "stereospecificity" of a catalyst will be polymerization of propylene to an isotactic polymer.

yield isotactic polypropylene under conventional conditions of polymerization. This is probably due to the formation of the dimeric complex:

$$2Al(C_2H_5)_2OC_6H_5 \rightleftarrows (C_2H_5)_2Al\begin{matrix} C_6H_5 \\ | \\ O \\ \diagup \quad \diagdown \\ \diagdown \quad \diagup \\ O \\ | \\ C_6H_5 \end{matrix}Al(C_2H_5)_2$$

These complexes are fairly stable at low temperatures, but are decomposed at elevated temperatures, with the result that they become catalytically active at about 130°C. The dimeric complexes are capable of reacting with Lewis bases:

$$Al_2(C_2H_5)_4X_2 + 2D \longrightarrow 2D \cdot Al(C_2H_5)_2X$$

where D is a Lewis base.

The following sequence of various X groups reflects their relative tendency to complex formation:

$$C_2H_5 < I < Br < Cl < F < SeC_6H_5 < SC_6H_5 < OC_6H_5 < NC_5H_{10}$$

The second group of organoaluminum compounds which may be constituents of catalytic systems includes a number of complexes of alkylaluminum compounds with Lewis bases $D \cdot Al(C_2H_5)_3$, where D is a Lewis base of the type $(C_2H_5)_2O$, $N(CH_3)_3$, etc. Depending on the strength of the Lewis base, the activity of the catalyst varies. If very strong Lewis bases such as pyridine or trimethylamine are employed, the active catalyst is not formed, probably owing to the formation of stable donor-acceptor complexes of the type $D \cdot AlR_3$. Weaker Lewis bases such as diethyl ether react with titanium trichloride to form a stereospecific catalyst. According to other observations /12/, Lewis bases may be employed not only to adjust the stereospecificity of the catalyst, but also the molecular weight of the resulting polymer.

The third group of organoaluminum compounds includes complexes of alkylaluminum compounds with metal salts in conjunction with a Lewis base, e.g., $KCl \cdot Al(C_2H_5)_3 + (C_2H_5)_2O$. This complex together with titanium trichloride forms an effective catalyst, yielding a polypropylene which is about 80% isotactic. However, some salts of similar nature, such as sodium fluoride, do not form an active catalyst; the reasons for this are not clear.

The advantage of using this type of organoaluminum compounds as component of a catalyst complex with transition metal salt is that these compounds are much less flammable than pure alkylaluminum compounds and are thus easier to handle.

The fourth group of the catalyst systems includes organoaluminum compounds of the type $Al(C_2H_5)X_2$ in conjunction with a Lewis base. The compound reacts with a Lewis base to form a very active stereospecific catalyst:

$$2Al(C_2H_5)Cl_2 + D \longrightarrow AlCl_3 \cdot D + Al(C_2H_5)_2Cl$$

Polypropylene which has been polymerized by this catalyst is up to 98% isotactic. The active organoaluminum compound proper is $Al(C_2H_5)_2Cl$, i.e., a compound of the first group. However, the compound is formed in situ and is therefore somewhat more effective than a previously prepared AlR_2Cl.

It thus follows from the results obtained by Natta and his school that the nature of the substituent in the organoaluminum compound affects not merely the activity (polymerization rate), but also the stereospecificity of the catalyst complex.

2. EFFECT OF THE TRANSITION METAL IN THE COMPLEX CATALYST ON ITS ACTIVITY AND STEREOSPECIFICITY

It may now be considered as conclusively established /9, 48/ that the identity of the transition metal which participates in the formation of the catalyst complex is of paramount importance in the course of the growth of the polymer chain, i.e., in the formation of a definite microstructure.* This was strikingly shown by a comparative study of homogeneous and heterogeneous catalyst complexes.

As is well known, compounds of transition metals with unfilled d-orbitals form complexes with olefins, which catalyze the polymerization of vinyl monomers. The filled π-orbitals of the olefin overlap the d-orbitals of the metal; conversely, the filled d-orbitals of the metal also overlap the empty antibonding orbitals of the olefin /13/.

The double bond in the π-complexes of transition metals with olefins are longer and thus also more active. Thus, the distance C — C in the complex $[Pt(C_2H_4)_2NH(CH_3)_2]Cl_2$ is 1.47 Å, while in the free olefin molecule it is 1.34 Å /14/.

Depending on the kind of metal orbitals which are overlapped by the π-orbitals of the olefin and on the extent of this overlapping, the resulting complexes of transition metal with the olefin may have different structures and thus also different reactivities. It follows that the structure and the properties of the polymer product will mainly depend on the type of the transition metal compound which constitutes the complex. The data given below indicate that there is a certain connection between the original form of the transition metal in the catalyst complex used in the polymerization and the structure of the polymer product /11/:

Original form of transition metal in the catalyst complex	Main type of stereoisomer in polymer
1. Valency less than the highest valency. Salt insoluble in reaction medium: $TiCl_2$, $TiCl_3$, VCl_3, etc.	Isotactic
2. Highest valency. Salt soluble in reaction medium: $TiCl_4$, VCl_4, etc.	Stereoblock
3. Oxidized form: $TiCl(OR)_3$, $Ti(OR)_4$, etc.	Atactic

* The microstructure of a polymer chain is understood to mean the mutual location of its constituent atoms and groupings.

A comparative study of heterogeneous (i. e., forming a solid salt of transition metal, e. g., $TiCl_3$ during complex formation) and homogeneous (fully soluble in the reaction mixture) complex organometallic polymerization catalysts clearly showed that the presence of the solid phase often does not affect the chain structure or other properties of the resulting polymer at all. Thus, polymerization of ethylene in the presence of a soluble complex catalyst $Sn(C_6H_5)_4 + AlBr_3 + VCl_4$ /15, 16/ yielded a high-molecular, linear polyethylene, with a melting point of 144°C (the same as the melting point of the polymethylene obtained from diazomethane), with a very narrow molecular weight distribution. Another, heterogeneous complex catalyst /15/

$$VCl_4 + C_2H_5-Al\left\langle\begin{array}{l} N\left\langle\begin{array}{l}Si(CH_3)_3\\ Si(CH_3)_3\end{array}\right. \\ N\left\langle\begin{array}{l}Si(CH_3)_3\\ Si(CH_3)_3\end{array}\right.\end{array}\right.$$

yielded an identical polyethylene, m. p. 144°C. This polymer was also distinguished by its uniform molecular weight.

On the other hand, polyethylene prepared in the presence of the system

$$TiCl_4 + C_2H_5-Al\left\langle\begin{array}{l} N\left\langle\begin{array}{l}Si(CH_3)_3\\ Si(CH_3)_3\end{array}\right. \\ N\left\langle\begin{array}{l}Si(CH_3)_3\\ Si(CH_3)_3\end{array}\right.\end{array}\right.$$

displayed the chemical structure and the thermomechanical behavior of an ordinary Ziegler polymer /16, 45/.

A similar effect was noted in the study of the copolymerization of ethylene and propylene on homogeneous and heterogeneous catalyst complexes. The copolymers had a more uniform structure and molecular weight if prepared in the presence of vanadium rather than titanium, irrespective of whether the catalyst complex was homogeneous or heterogeneous /17, 18/.

The above rule remains valid for the polymerization of butadiene as well, as may be seen from the following examples.

A study of two types of cobalt catalysts — the heterogeneous system $Al(C_2H_5)_2Cl + CoCl_2$ and the soluble system $Al(C_2H_5)_2Cl + CoCl_2 \cdot$ pyridine /19, 20/ — showed that the presence of a solid surface in the system does not affect the stereospecificity of the catalyst during the polymerization of butadiene. The polymerization of butadiene in a homogeneous system at 5°C was very rapid and the conversion was 100%. The resulting butadiene contained (in %):

1,4-links with cis-configuration	up to 98
1,4-links with trans-configuration	1 – 2
1,2-links	1 – 2

The heterogeneous catalyst was not as active, but the microtacticity of the resulting polymer was exactly the same.

Many more examples of polymerizations of olefins and dienes in the presence of complex organometallic catalysts could be quoted; they all indicate that the action of homogeneous and heterogeneous catalyst systems is the same for a given transition metal constituent.

However, phase relationships in the catalyst are a very relevant factor in the polymerization of α-olefins. All attempts to prepare isotactic polypropylene in the presence of homogeneous catalyst systems have so far failed. On the contrary, it was found that certain homogeneous systems (of the bis-cyclopentadiene type), which catalyze the polymerization of ethylene, are practically inactive with respect to other α-olefins /47/.

3. POLYMERIZATION OF ETHYLENE

Two types of organometallic catalysts are employed in the industrial polymerization of ethylene to the high-molecular, technically valuable polyethylene.

1. Complexes formed by the reaction between AlR_3 and $TiCl_4$. These are heterogeneous, i.e., are insoluble in the reaction medium owing to the reduction of the titanium to trivalent titanium and possibly also to lower valencies.

2. Soluble catalyst complexes, mostly prepared from bis-cyclopentadienyltitanium dichloride $(C_5H_5)_2TiCl_2$ and organoaluminum compounds.

Both catalyst types are suitable for polymerizing ethylene in a hydrocarbon solvent (hexane, or simply purified gasoline), at 50 — 70°C and under atmospheric or slightly elevated pressure.

The heterogeneous complexes $AlR_3 + TiCl_4$ are the most readily available and the simplest to prepare. This is the catalyst used by most industrial undertakings manufacturing low-pressure polyethylene /21/.

The overall content of the components of the heterogeneous catalyst system $AlR_3 + TiCl_4$ is usually less than 1% of the medium in which polymerization is carried out. Under optimum conditions of the polymerization, in which the purity of the starting ethylene is not less than 99.9%, and the system being practically free from moisture, sulfurous compounds and other harmful substances, the consumption of the catalyst is very low and the yield of the polyethylene is more than 1 kg per 1 gram of the catalyst.

The use of the theoretically and practically attractive soluble catalyst systems reduces the catalyst consumption even further. The study of heterogeneous catalyst systems is difficult owing to the need for preparing catalyst samples with preset (reproducible) lattice properties of the solid phase and particle size.

It may be expected that in soluble catalyst systems each ionic bond can react with the monomer and initiate a polymer chain growth. As a result, the consumption of a soluble catalyst will be much smaller than that of a heterogeneous catalyst, which has only a small proportion of active sites on its surface. From the technological point of view, homogeneous catalysts are much to be preferred, since when these are employed, it is much easier to remove the residual catalyst from the polymer. Natta et al. /22/ and Breslow /23/ were the first to note that high-molecular polymers are formed when ethylene is polymerized in the presence of soluble catalyst complexes consisting of bis-cyclopentadienyltitanium dichloride $(C_5H_5)_2TiCl_2$

and an organoaluminum compound. They showed that when $(C_5H_5)_2TiCl_2$ is made to react with $Al(C_2H_5)_3$, a "sandwich" structure of the formula $(C_5H_5)_2TiCl_2Al(C_2H_5)_2$ is formed, which then further reacts with $Al(C_2H_5)_3$.

Shilov et al. /24 — 26, 47/ studied the mechanism of the polymerization of ethylene in the presence of complexes of this type and showed that the mechanism is very probably ionic. Complexes containing pentadienyl compounds of metals other than titanium also proved active /27/. Table 11 illustrates the differences in the properties of polyethylene prepared in the presence of the soluble bis-cyclopentadienyl complex and in the presence of the conventional Ziegler catalyst.

TABLE 11. Properties of polyethylenes prepared in the presence of complex catalysts of different types

Parameter	Homogeneous bis-cyclopentadienyl complex	Ziegler catalyst $AlR_3 + TiCl_4$
Intrinsic viscosity	2.3	2.3
M.p., °C	137	132
Crystallinity, %	85	79
Functional group content, %:		
methyl	0.05	0.86
vinyl	0.036	0.044
vinylidene	0.006	0.007
Density at 25°C, g/cm^3	0.951	0.945

Oxygen has an activating effect in the polymerization of ethylene in the presence of bis-cyclopentadienyl compounds: oxygen present in very small amounts (0.025 mole %) resulted in a considerably increased catalytic activity. The part played by oxygen was clarified in more detail, in the case of polymerization of ethylene in the presence of another catalytic system on the base of tetraphenyltin /28 — 30/. This catalyst is formed on mixing together definite proportions of $AlCl_3$ or $AlBr_3$, tetraphenyltin and a small amount of a halide of vanadium (VCl_4 or $VOCl_3$) in cyclohexane or in another hydrocarbon.

The polymerization of ethylene in the presence of this catalyst has been studied in detail /15, 16, 31/. The main component of this catalyst — tetraphenyltin — is stable to oxygen and to water, unlike several other organometallic compounds, in particular alkylaluminum compounds. The system forms a true solution, as indicated by the absence of Tyndall effect and by the fact that its catalytic effect is unimpaired after filtration through a bacterial filter with 1μ pore diameter.

It has been shown that the first stage in the polymerization of ethylene in the presence of such a system is the reaction of tetraphenyltin with aluminum tribromide:

$$Sn(C_6H_5)_4 + 2AlBr_3 \longrightarrow 2AlC_6H_5Br_2 + Sn(C_6H_5)_2Br_2$$

The second stage consists in the formation of an active catalyst complex between $AlC_6H_5Br_2$ and VCl_4; the transition metal supplies the vacant d-orbitals for the formation of the π-complex with the olefin.

The introduction of the first monomer molecule and the formation of the active complex constitutes the act of initiation of the polymerization (Chapter II). The entry of the following monomer molecules (propagation of polymer chain) takes place by attack on the V — C bond, which is weak, as in all highly alkylated organoaluminum compounds. The role played by the "aluminum" part of the active complex, which is a typical Lewis acid, is probably to produce a preliminary coordination of the monomer molecules which facilitates their subsequent attack on the bond between carbon and the transition metal.

The structure of the polyethylene obtained in the presence of the soluble complex differs from that of the polyethylene prepared using the ordinary Ziegler complex; the former is strictly linear (Table 12).

TABLE 12. Structure of polyethylene prepared in the presence of homogeneous and heterogeneous catalyst systems

Groups	In polyethylene prepared in the presence of soluble catalyst system $Sn(C_6H_5)_4 + AlBr_3 + VCl_4$	In polyethylene prepared in the presence of heterogeneous catalyst $Al(C_2H_5)_3 + TiCl_4$
Methyl	0.05	0.5
Vinyl	0.01	0.07
Vinylidene	0.003	0.008

The catalyst system $Sn(C_6H_5)_4 + AlBr_3 + VCl_4$, which makes it possible to conduct polymerizations of ethylene at a rapid rate, soon loses its activity. Attempts were accordingly made to find another soluble catalyst system, which would ensure a constant reaction rate for long periods of time. The organometallic constituent of such a complex was the hitherto unknown compound, which is a modified alkylaluminum compound containing both Al — C and Al — N bonds. This compound is a weak reducing agent with respect to the transition metal salt, in particular $TiCl_4$, which ensures the homogeneity of the system, and is at the same time sufficiently active to form a highly effective catalytic coordination complex. This complex ensures a prolonged polymerization of ethylene, which proceeds at a practically constant rate /45/.

Other soluble organometallic complex systems, which catalyze the polymerization of ethylene, are also available.

A complex catalyst of polymerization of ethylene, which is soluble in chlorinated hydrocarbons, is formed by the reaction between VCl_4 and $Al(iso\text{-}C_4H_9)_2Cl$ /32/. The molecular weight of the polyethylene formed in its presence is high — about 10^6. Such a polyethylene is difficult to work, but has certain industrial uses — in particular, the manufacture of loom shuttles — owing to its hardness. Studies of soluble complex polymerization catalysts are continuing.

4. STEREOSPECIFIC POLYMERIZATION OF PROPYLENE AND OTHER UNSATURATED HYDROCARBONS

Advances in the synthesis and in the various applications of complex organometallic catalysts are largely connected with the realization of stereospecific polymerization of propylene and other unsaturated hydrocarbons, which has now become an industrial process practised in many countries.

The principle of stereospecific polymerization is the selection of certain configurations of monomeric links during the polymerization process. If this selection can be controlled, the resulting polymers will be stereoregular, i.e., will have an ordered structure; they will therefore be capable of crystallization and will display a number of technically valuable properties such as a high strength, ready fiber formation, etc.

If the polymerization is to be stereospecific, the monomer molecules about to participate in the polymerization must be preliminarily ordered. This is best achieved with the aid of organometallic complex catalysts which include compounds of transition metals. It has already been said that it is the bond between the carbon atom and the transition metal which is attacked.

The simplest stereoregular polymer is polypropylene, the macromolecules of which can be isotactic or syndiotactic.

The two main problems in this field are the study of the formation of isotactic polypropylene and the development of a process for its industrial manufacture.

Best results were obtained in preparing isotactic polypropylene by using the complex catalyst system $AlR_2X + TiCl_3$, where X is an atom of halogen or an ethyl group. This method was first proposed and studied in detail by Natta /8/, and the catalyst has now received general acceptance in the production of polypropylene.

The form of $TiCl_3$ crystals has an important effect on the stereospecific activity of the system. As is well known, four crystalline modifications of this salt are known — the α-, β-, γ- and δ-forms. The brown β-$TiCl_3$ is the least stereospecific, while the activities of the violet α-, γ- and δ-forms are about the same.

TABLE 13. Stereospecificity of systems catalyzing the polymerization of propylene

Organometallic compound	Amount of isotactic polymer formed		Organometallic compound	Amount of isotactic polymer formed	
	at 15° C	at 70° C		at 15° C	at 70° C
$Al(C_2H_5)_2I$	99 — 100	96 — 98	$Al(C_2H_5)_3$	80 — 85	80 — 85
$Al(C_2H_5)_2Br$	87 — 98	94 — 96	$Be(C_2H_5)_3$	94 — 96	93 — 95
$Al(C_2H_5)_2Cl$	96 — 98	91 — 94			

Table 13 shows /11/ the stereospecific effects of the active (violet) forms of $TiCl_3$ in the catalyst systems on the polymerization of propylene. It follows from the table that systems containing $Al(C_2H_5)_2I$ are the most effective; however, $Al(C_2H_5)_2Cl$ is more readily available and its stereospecificity is also satisfactory, so that it is used instead. Alkylberyllium compounds are not employed on account of their toxicity.

It was repeatedly attempted to use catalyst complexes containing salts of transition metals other than titanium. Some of the results are shown in Table 14. It follows from these data that optimum conditions for the polymerization of propylene are ensured by employing complexes with lower valency titanium and vanadium chlorides.

TABLE 14. Yield of crystalline polypropylene as a function of the type of transition metal compound in the complex with AlR_3 /11/

Transition metal compound	Yield of crystalline polypropylene, %	Transition metal compound	Yield of crystalline polypropylene, %
$TiCl_2$	80 — 90	$TiCl_4$	45 — 50
$TiCl_3$		$TiBr_4$	40 — 42
α-form	85 — 90	TiI_4	46
β-form	40 — 50	$Ti(OR)_4$	Traces
$TiBr_3$	44	$ZrCl_4$	52
TiI_3	10	VCl_4	48
$ZrCl_2$	55	$VOCl_3$	32
VCl_3	70 — 75		
$CrCl_3$	36		

Like polyethylene, polypropylene too may be prepared in the presence of soluble catalyst complexes. It must be remembered that, in the case of the heterogeneous catalyst systems described above, it is only a small proportion of the ionic bonds in the complex — those on the surface of the solid phase — that display catalytic activity. For example, the number of active (polymerization-promoting) sites in $TiCl_3$ specimens used in the stereospecific polymerization of propylene is only a few units per 1000 $TiCl_3$ molecules. Such calculations are effected with the aid of alkylaluminum compounds tagged with radioactive carbon. In a soluble catalyst complex, on the contrary, each ionic bond is free to react with the monomer and thus initiate polymer chain propagation.

However, stereospecific polymerization of propylene with the aid of soluble catalyst systems involves serious difficulties. In order for the course of a polymerization to be stereospecifically controlled, the monomer must be bound to the catalyst complex in a definite manner, i.e., the resulting complex between the catalyst and the monomer must have a definite spatial configuration. Such complexes may be formed by monomers which contain polar groups and by diene hydrocarbons with a second double bond, which then acts as a polar substituent. This complex formation cannot be realized if propylene is polymerized under ordinary conditions in the presence of soluble catalysts.

However, at low temperatures, when the molecular motion of the monomer is limited, propylene may be stereospecifically polymerized with formation of syndiotactic structure.

The soluble catalyst systems employed for the purpose at −78°C included complexes between vanadium acetylacetonate or VCl_4 and anisole and AlR_2Cl or AlR_2F. The yield of the syndiotactic polypropylene was very low /33/. Subsequent studies also failed to give a significant increase in the yield of this polymer. Soluble catalyst systems have not so far found use in industrial scale polymerization of propylene, but research on the subject is continuing.

However, soluble complex catalyst systems proved effective in copolymerizations of propylene and ethylene to elastomers. A number of catalyst systems, some of which are industrially employed, have been proposed for the copolymerization of ethylene and propylene to an amorphous rubber-like product /34 — 38/. Most of these systems contain vanadium compounds in the capacity of transition metal compound: $VO(C_2H_5)_3$, $VO(C_2H_5)_2Cl$, vanadium acetylacetonate $V(C_5H_7O_2)_3$ or vanadyl acetylacetonate $VO(C_5H_7O_2)_2$.

Experimental results showed that copolymerization of ethylene with propylene in all cases resulted in a copolymer of a homogeneous structure. The molecular weight distribution was narrow, as in the homopolymerization of ethylene.

Soluble organometallic complexes are successfully employed in stereospecific polymerization of diene hydrocarbons — butadiene and isoprene. In this way stereoregular rubbers are obtained, of a quality comparable to that of natural rubber. However, when dienes are polymerized in the presence of organometallic catalyst complexes, a number of special problems arise, the discussion of which is outside the scope of this book; we shall merely mention that very interesting theoretical and practical data were obtained in the course of the past few years /46/ on stereospecific polymerization of butadiene in the presence of organometallic catalysts based on π-allyl complexes of a number of transition metals.

Organometallic catalyst complexes have also found use in the polymerization of α-olefins other than propylene: butene-1, 3-methylbutene-1, 4-methylpentane-1 /21/. It was found that in the presence of complexes of the type $AlR_3 + TiCl_4$ (or $AlR_3 + TiCl_3$) unbranched α-olefins with 5 carbon atoms or more polymerize to give low-melting, mostly amorphous, low-molecular substances of little practical interest.

If branched α-olefins are polymerized under similar conditions, the resulting polymers are high-molecular, stereoregular and crystalline, and have a higher melting point than polypropylene (Table 15), which is a desirable property.

Polymers such as poly(3-methylbutene-1) and poly(4-methylpentene-1) have found industrial utilization in the production of high strength, heat resistant polyhydrocarbon fibers. Heat-resistant organic glasses based on poly(4-methylpentene-1) are known.

TABLE 15. Melting points of polymers prepared from different α-olefins in the presence of the complex catalyst $AlR_3 + TiCl_4$

Initial monomer	M.p., °C	Initial monomer	M.p., °C
Propylene	165 — 170	3-Methylbutene-1	250
Butene-1	128	4-Methylpentene-1	230 — 240
Pentene-1	80	4-Methylhexene-1	170 — 180

Vinylcyclohexane may be polymerized in the presence of complex organometallic catalysts to yield the high-melting (above 300°C) polyvinylcyclohexane /51/.

Organometallic complexes also catalyze the polymerization of acetylenic hydrocarbons — acetylene and its derivatives. The resulting polymers contain conjugated systems of C = C bonds. The simplest such polymer is the product of polymerization of acetylene itself:

$$nHC{\equiv}CH \longrightarrow [-CH{=}CH-CH{=}CH-CH{=}CH-]_n$$

Polymers with a system of conjugated double bonds display a number of unusual properties; in particular, they are semiconductors /39/.

Complex catalysts of the type $AlR_3 + TiCl_4$ or similar systems readily catalyze the polymerization of acetylene to a black-colored, polyconjugated compound. In this polymerization, the active catalyst species is formed by the reaction between $Al(C_2H_5)_3$ and vanadyl acetylacetonate /40/:

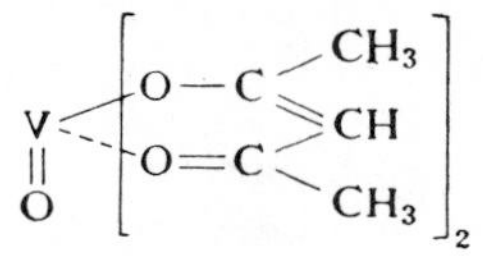

$$O{=}V\left[\begin{matrix} O-C\langle^{CH_3}_{\geq CH} \\ O{=}C\langle_{CH_3} \end{matrix}\right]_2$$

Our own investigations showed that this catalyst system is most active if the molar ratio Al:V = 4. As triethylaluminum was replaced by $Al(C_2H_5)_2Cl$ and then by $Al(C_2H_5)Cl_2$, the catalytic effect in the polymerization of acetylene strongly decreased.

Catalysts based on triethylaluminum and titanium tetrachloride are active in the polymerization of unconjugated diynes as well: 1, 6-heptadiyne, 1, 7-octadiyne and 1, 8-nonadiyne /41/. The polymers contain a conjugated system and rings in the chain:

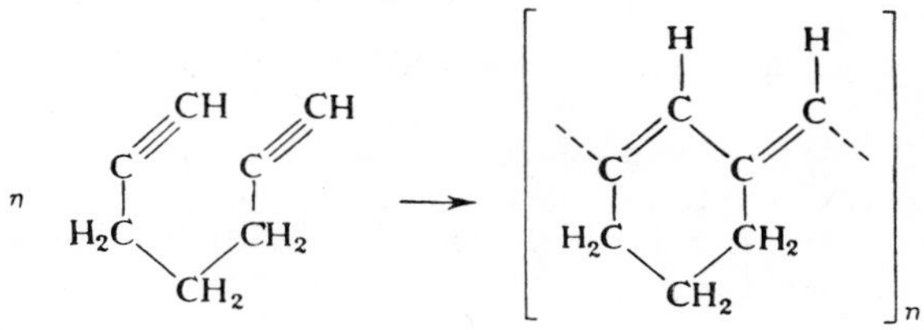

$$n\ HC{\equiv}C{-}CH_2{-}CH_2{-}CH_2{-}C{\equiv}CH \longrightarrow \left[\begin{matrix} H & & H \\ | & & | \\ \cdots C{=}C & -C{=}C\cdots \\ | & & | \\ H_2C & -CH_2- & CH_2 \end{matrix}\right]_n$$

The polymerization of various classes of hydrocarbons with the aid of organometallic complex catalysts has formed the subject of numerous studies. These studies are still continuing and the catalytic potentialities of the different organometallic systems are still far from exhausted.

5. UTILIZATION OF COMPLEX ORGANOMETALLIC CATALYSTS IN THE POLYMERIZATION OF NONHYDROCARBON MONOMERS

Immediately following the discovery and utilization of complex organometallic catalysts in polymerization reactions it was believed that they could not be utilized in the polymerization of nonhydrocarbon monomers. However, subsequent studies showed that the new catalysts can also be used to promote stereoregular polymerizations of a wide range of monomer compounds, i.e., vinyl chloride /42/.

Poly(vinyl chloride) is even now the most common thermoplastic material, both as regards its production volume and its variety of applications. It is thus natural that repeated attempts are being made to improve its properties, in particular its resistance to heat and its mechanical parameters. It is believed that one way to attain this objective is to produce crystalline, stereoregular poly(vinyl chloride). Ordinary catalysts of stereospecific polymerization, based on titanium chlorides and alkylaluminum compounds, cannot be employed in this case, since monomers with active halogens are usually decomposed by such catalyst systems. It was recently shown however /42a/, that if titanium fluoride is used instead of titanium chloride, the catalyst system $TiF_4 + Al(iso\text{-}C_4H_9)_3$ will produce a stereoregular poly(vinyl chloride), though in a low yield.

Crystalline poly(vinyl chloride) can also be prepared using the catalyst $Al(C_2H_5)_3 + TiCl_3$ in a saturated hydrocarbon medium, in the presence of special nucleophilic complexants as additives. It must be noted, however, that the literature data concerning the degree and type of stereoregular structure of the polymer are based on IR spectroscopic data only and require confirmation.

Complex organometallic catalysts are also successfully employed in stereospecific polymerizations of vinyl ethers $ROCH = CH_2$. Catalysts which can be employed in this way include both Ziegler type catalysts and a number of catalysts based on boron fluorides and alkylaluminum compounds, which act by a cationic mechanism /43/. These catalysts — the so-called Nakano catalysts — appear to form complex ions, which are capable of acting as multiple site catalysts:

$$BF_3 + R_3Al \longrightarrow R_2Al^+ + [BF_3 \cdot R]^-$$

$$[BF_3 \cdot R]^- + R_3Al \longrightarrow R_2AlR \cdots [BF_3R]^-$$

The principal features of stereospecific polymerization of several vinyl ethers have been studied in fair detail.

Complexes of AlR_3 with salts of transition metals are also active in the polymerization of alkylene oxides $\underset{\diagdown O \diagup}{RCH - CH_2}$ when the oxygen ring is opened and polymers of the type:

$$-\underset{\displaystyle R}{\underset{|}{CH}}-CH_2-O-\underset{\displaystyle R}{\underset{|}{CH}}-O-CH_2-$$

are formed.

Furukawa and Saegusa /44/ recently conducted interesting studies on the synthesis of polymers by the polymerization of polar monomers such as aldehydes, olefin oxides, etc. They showed that one of the most common industrial chemicals — acetaldehyde — can be polymerized to a crystalline polyacetaldehyde in the presence of organometallic compounds, mainly triethylamine in conjunction with a number of metal salts, water, alcohol and other compounds. A study of the polymer yield as a function of the composition of aqueous triethylaluminum showed that the compound $(C_2H_5)_2Al-O-Al(C_2H_5)_2$ is the most active. Catalysts of this type proved to be active in polymerization reactions of a large number of polar monomers. A recent achievement is the polymerization of other polar monomers such as β-isovalerolactam /44a/, tetrahydrofuran /49/ or methyl methacrylate /50/ in the presence of catalysts consisting of complex organometallic compounds prepared from alkylaluminum compounds and salts of transition metals.

Thus, the original scope of application of complex organometallic catalysts in macromolecular chemistry has now been much extended. This notwithstanding, their most important application is still the stereospecific polymerization of unsaturated hydrocarbons of different classes; this is at the same time their most extensive and most important industrial application.

Bibliography

1. Gaylord, N. and H.F. Mark. Linear and Stereoregular Addition Polymers. — New York, Interscience. 1959.
2. Mosevitskii, M.I. — Uspekhi Khimii, **28**:466. 1959.
3. Topchiev, A.V., B.A. Krentsel', and L.L. Stotskaya. — Uspekhi Khimii, **30**:462. 1962.
4. Kuper, G.M. — Khimiya i Tekhnologiya Polimerov, **10**:3. 1962.
5. Topchiev, A.V., L.L. Stotskaya, and B.A. Krentsel'. — Plasticheskie Massy, **12**:3. 1962.
6. Natta, G. and J. Pasquon. — Kinetika i Kataliz, **3**:805. 1962.
7. Bawn, C. and A. Ledwith. — Quart. Rev., **16**:361. 1962.
8. Natta, G. — Khimiya i Tekhnologiya Polimerov, **1**:42. 1965.
9. Baranova, G.A., L.L. Stotskaya, and B.A. Krentsel'. — Kinetika i Kataliz, **1**:24. 1967.
10. Pasquon, J., A. Zambelli, and G. Gati. — Makromol. Chem., **61**:116. 1963.
11. Krentsel', B.A. and L.G. Sidorova. — Polipropilen (Polypropylene). Kiev, Izd. "Tekhnika." 1964.
12. Minsker, K.S. Author's Summary of Doctoral Thesis. FKhI im. Karpova. Moskva. 1964.
13. Cossee, P. — J. of Catalysis, **3**:80. 1964.
14. Alderman, R., P. Owston, and J. Rowe. — Acta cryst., **13**:149. 1960.
15. Stotskaya, L.L. Author's Summary of Thesis. INKhS AN SSSR, Moskva. 1963.
16. Stotskaya, L.L. and B.A. Krentsel'. — Doklady AN SSSR, **151**:595. 1963.
17. Bier, G., A. Gumboldt, and G. Schleitzer. — Makromol. Chem., **58**:43. 1962.
18. Dall'asta, G., G. Mazzanti, G. Natta, and L. Porri. — Makromol. Chem., **56**:224. 1962.
19. Zgonnik, V.N., B.A. Dolgoplosk, N.I. Nikolaev, and V.A. Kropachev. — Vysokomolekulyarnye Soedineniya, **4**:1000. 1962.
20. Dolgoplosk, B.A., E.N. Kropacheva, E.K. Khrennikova, E.I. Kuznetsova, and K.G. Golodova. — Doklady AN SSSR, **135**:847. 1960.
21. Topchiev, A.V. and B.A. Krentsel'. Poliolefiny novye sinteticheskie materialy (Polyolefins as New Synthetic Materials). — Izd. "Nauka." 1963.
22. Natta, G., P. Pino, and G. Mazzanti. — J. Am. Chem. Soc., **79**:2975. 1957.
23. Breslow, D. and N. Newburg. — J. Am. Chem. Soc., **79**:5072. 1957; **81**:81. 1959.

24. Zefirova, A.K. and A.E.Shilov. — Doklady AN SSSR, **136**:599. 1961.
25. Zefirova, A.K., N.I.Tikhomirova, and A.E.Shilov. — Doklady AN SSSR, **132**:1082. 1960.
26. Shilov, A.E. Author's Summary of Thesis, Institute Khim. Fiziki AN SSSR, Moskva. 1966.
27. Kaar, Kh. Author's Summary of Thesis. INKhS AN SSSR, Moskva. 1964.
28. Carrick, W. — J.Am.Chem.Soc., **80**:6455. 1958.
29. Carrick, W. et al. — J.Am.Chem.Soc., **82**:3883. 1960.
30. Phillips, G. and W.Carrick. — Khimiya i Tekhnologiya Polimerov, **11**:3. 1961.
31. Stotskaya, L.L., I.F.Leshcheva, and B.A.Krentsel'. — Neftekhimiya, **4**(1):33. 1964.
32. Zavorokhin, N.D. and A.I.Makhan'ko. — Vestnik AN KazSSR, **6**:19. 1966.
33. Natta, G., I.Pasquon, and A.Zambelli. — J.Am.Chem.Soc., **84**:1488. 1962.
34. Carrick, W. — J.Am.Chem.Soc., **82**:1502. 1960.
35. Karapinka, G., J.Smith, and W.Carrick. — J.Polymer Sci., **50**:143. 1961.
36. Natta, G. and G.Mazzanti. — J.Polymer Sci., **51**:441. 1961.
37. Natta, G. et al. — J.Polymer Sci., **34**:88. 1959.
38. Phillips, G. and W.Carrick. — J.Am.Chem.Soc., **84**:920. 1962.
39. Topchiev, A.V. (editor). — In: "Organicheskie poluprovodniki." Izd. "Nauka." 1962.
40. Nasirov, F.M. Author's Summary of Thesis. INKhS AN SSSR, Moskva. 1965.
41. Stille, S. and D.Frey. — J.Am.Chem.Soc., **83**:1697. 1961.
42. Krentsel', B.A. — Khimicheskaya Promyshlennost', **12**:865. 1962.
43. Collection "Uspekhi khimii polimerov." — Izd. "Khimiya." 1966.
44. Furukawa, J. and T.Saegusa. Polymerization of Aldehydes and Oxides. — New York, Interscience. 1963.
44a. Ciaperoni et al. — J.Polymer Sci., A-1, **5**(4):491. 1967.
45. Baranov, G.A., N.N.Korneev, B.A.Krentsel', and L.L. Stotskaya. — Vysokomolekulyarnye Soedineniya, **6**:1263. 1967.
46. Dolgoplosk, B.A. et al. Mekhanizm polimerizatsii dienov na π-allil'nykh kompleksakh (Mechanism of Diene Polymerization on π-Allyl Complexes). — Izd. "Nauka." 1968.
47. D'yachkovskii, F.S. and A.E.Shilov. — ZhFKh, **41**:2515. 1967.
48. Natta, G., L.Porri, and S.Valenti. — Makromol. Chem., **67**:225. 1963.
49. Gorin, Yu.A. et al. — Soviet Patent 188670. 1966.
50. Koide, Naoyuki et al. — Kogyo Kagaku Zasshi, **70**(7):1224. 1967.
51. Kleiner, V.I., L.L.Stotskaya, and B.A.Krentsel'. — Plasticheskie Massy, No. 4:3. 1967.

Chapter VI

OTHER UTILIZATIONS OF COMPLEX CATALYSTS

In addition to serving as catalysts of the polymerization of olefins, dienic hydrocarbons and other organic monomers, complex organometallic compounds also catalyze dimerization, oligomerization and ring closure of various hydrocarbons. They have been recently employed to catalyze the hydrogenation, isomerization and alkylation of numerous aliphatic and aromatic compounds. They have also served, in toto or as individual components of the complex, to catalyze the preparation of metal carbonyls and π-complexes of transition metals, and in the chemical fixation of molecular nitrogen. All these studies are of high interest and deserve to be discussed in detail.

1. PREPARATION OF LOW-MOLECULAR OLIGOMERS OF UNSATURATED HYDROCARBONS

Dimerization and oligomerization of olefins

Ziegler et al. /1/ were the first to report that the reaction of triethylaluminum with ethylene in the presence of colloidal nickel results in the formation of butene-1. Cobalt and platinum act in a similar manner. The process mechanism is catalytic between 100 and 120°C under 40 — 60 atm pressure and the yield of butene-1 is high (more than 90%). It has been recommended that a small amount of an acetylenic hydrocarbon be added to the catalyst in order to ensure a steady rate of reaction. For a review of earlier publications on the dimerization of ethylene see /2, 3/.

A number of effective catalysts have been recently proposed. A system with cobalt acetate or cobalt acetylacetonate as one component and triethylaluminum, diisobutylaluminum hydride, diethylaluminum ethoxide or sodium aluminum tetraethyl in benzene, heptane or tetrahydrofuran as the other, will dimerize ethylene even at 0°C /4/. Catalyst obtained by treating nickel acetylacetonate with triphenylphosphine and a large excess of $(C_2H_5)_3Al_2Cl_3$ is extremely active /5/. Within 18 minutes $18.3 \cdot 10^5$ moles of ethylene are dimerized by 1 mole of $Ni(C_5H_7O_2)_2$, and the latter still remains fully active. The product contains 94% butylene (almost exclusively butene-2).

The dimerization of propylene is also of major practical interest. It is the result of a reaction between propylene and tripropylaluminum /6/. During the first stage an unstable alkylaluminum compound is formed:

$$CH_3-CH_2-CH_2-Al + C_3H_6 \longrightarrow \begin{array}{l} CH_3-CH_2-CH_2 \\ \quad\quad\quad\quad | \\ CH_3-CH-CH_2-Al \end{array}$$

It then reacts with propylene and is decomposed as follows:

$$\begin{array}{l} CH_3-CH_2-CH_2 \\ \quad\quad\quad\quad | \\ CH_3-CH-CH_2-Al \end{array} + C_3H_6 \longrightarrow \begin{array}{l} CH_3-CH_2-CH_2 \\ \quad\quad\quad\quad | \\ CH_3-CH=CH_2 \end{array} + \begin{array}{l} CH_3 \\ | \\ CH_2CH_2Al \end{array}$$

If the propylene dimer (2-methylpentene-1) is led out of the reaction mixture, the mechanism of the process becomes catalytic and a small amount of the alkylaluminum compound /7 — 9/ or of its complex with a nickel salt /9a, 9b/ will suffice to dimerize the propylene. The dimerization of propylene has lately attracted much attention in connection with a new method for the synthesis of isoprene, which has the preparation of 2-methylpentene-1 as its first stage. 2-Methylpentene-1 is then isomerized to 2-methylpentene-2, which gives high yields of isoprene by liberation of methane:

$$\begin{array}{l} CH_3-CH_2-CH_2 \\ \quad\quad\quad\quad | \\ CH_3-C=CH_2 \end{array} \longrightarrow \begin{array}{l} CH_3-CH_2-CH \\ \quad\quad\quad\quad \| \\ CH_3-C-CH_3 \end{array} \xrightarrow[HBr]{650-800^\circ C}$$

$$\longrightarrow \begin{array}{l} CH_2=CH \\ \quad\quad | \\ CH_3-C=CH_2 \end{array} + CH_4$$

The dimerization of ethylene in the presence of organoaluminum compounds may be considered as a special case of synthesis of higher olefins by the "growth" of alkylaluminum compounds and displacement of the olefin, when the rate of the latter process much exceeds the rate of the former:

$$\begin{array}{rl} AlC_2H_5 + C_2H_4 & \longrightarrow AlC_2H_4C_2H_5 \\ AlC_2H_4C_2H_5 + nC_2H_4 & \longrightarrow Al(C_2H_4)_{n+1}C_2H_5 \\ Al(C_2H_4)_{n+1}C_2H_5 & \longrightarrow AlH + CH_2=CH-(C_2H_4)_nC_2H_5 \\ AlH + C_2H_4 & \longrightarrow AlC_2H_5\text{, etc.} \end{array}$$

If the displacement rate is slower than the propagation rate, the reaction will yield alkylaluminum compounds with long-chain alkyl radicals, which can in turn be converted to higher alcohols by oxidation, to acids by carbonation or to higher olefins. For an account of the theoretical and practical aspects of the propagation reaction the reader is referred to /2, 3, 10/. Here we shall merely mention the very characteristic reaction of chain propagation on bifunctional organoaluminum compounds, owing to which higher α, ω-diolefins, diols and dibasic acids can be prepared. It was pointed out by Ziegler /10/ that many attempts were made to realize such reactions; they were unsuccessful, since the readily available diolefins with 4 — 6 carbon atoms (butadiene, diallyl) are unsuitable for the preparation of bifunctional alkylaluminum compounds. In particular, under these experimental conditions diallyl became converted to methylenecyclopentane /11/. α, ω-Diolefins with 8 — 10 carbon atoms have recently become available and these can be converted to bifunctional alkylaluminum compounds. Results of such studies have not been reported by Ziegler, but patent literature contains several references to this subject. Thus, a patent /12/ has been taken out for the preparation of diols with 8 — 16 carbon atoms starting from butadiene, ethylene and alkylaluminum compounds. Butadiene reacts with $(iso\text{-}C_4H_9)_2AlH$ at 90 — 100°C for 4 hours with formation of tributenylaluminum, which is then treated with triisobutylaluminum and heated at 130 — 150°C until the evolution of isobutylene has ceased. In the resulting compound tetramethylene chains are bound to two aluminum atoms. When this compound is reacted with ethylene at 80 — 90°C, the alkyl chains first increased in size to $C_8 - C_{16}$, after which oxidation was carried out.

In another patent /13/ α, ω-diolefins are prepared from ethylene and tris(3, 3-dimethylpentamethylenyl)dialuminum in toluene in the presence of 0.1 — 1.0% of Ni, Co or Pd compounds (e.g., nickel acetylacetonate) and phenylacetylene. The reaction is carried out at 70°C and 60 atm for 3 hours. The product is then treated with acidified methanol at −20°C and distilled. The main fraction (83% yield) distils over at 56 — 160°C / 100 mm, which corresponds to an average molecular weight of 201.

Natta and Miyake /14/ studied the reaction of chain growth on derivatives of 1, 5-dialuminopentane. They showed that ring formation was absent. The compounds found after decomposition of the reaction product included C_5, C_7, C_9, C_{11}, C_{13}, etc., hydrocarbons. The chain length distribution approached the Poisson distribution. It was shown that all Al — C bonds react with the same probability throughout the reaction period.

Linear and ring oligomerization of diolefins

Wilke /15/ in 1957 succeeded in preparing the ring trimer of butadiene, trans-trans-trans-cyclododecatriene-1, 5, 9 (CDT) from butadiene in the presence of the catalyst system $(C_2H_5)_2AlCl - TiCl_4$:

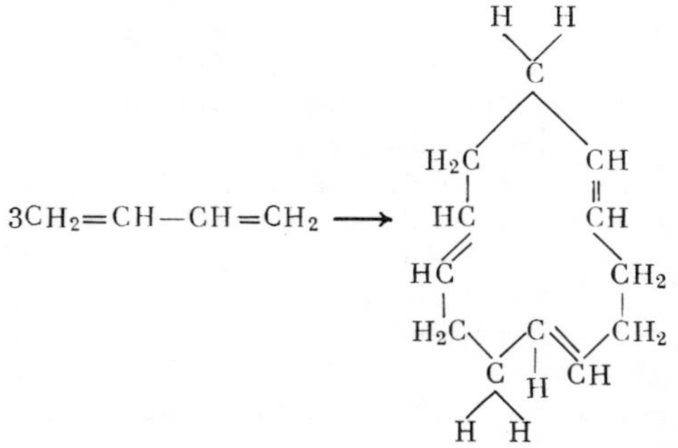

$$3CH_2{=}CH{-}CH{=}CH_2 \longrightarrow$$

H H
C
H_2C CH
HC CH
HC CH_2
H_2C C CH_2
C H CH
H H

The yield of the reaction product shows considerable variation with the Al:Ti ratio; the maximum CDT yield is obtained if the value of this ratio is 4.5, while if this ratio is 0.5 to 1.0, 1, 4-trans-polybutadiene is obtained. Other catalyst systems for the cyclotrimerization of butadiene were subsequently proposed /16/. Thus, CDT was prepared in 90 — 95% yield at 40°C under atmospheric pressure in the presence of a catalyst formed by mixing diethylaluminum hydride with $AlCl_3$ in benzene and adding $TiCl_4$ afterwards. High CDT yields were also obtained when the following catalysts were employed: $CaH_2 - AlCl_3 - TiCl_4$; $NiX_2 - AlR_3$ — electron donor (ether); $CrO_2Cl_2 — (C_2H_5)_3Al$. The last-named system gives an isomer mixture of CDT — trans-trans-trans-isomer and trans-trans-cis-isomer. Both CDT isomers were also prepared by passing butadiene through a benzene solution of $(C_2H_5)_2AlCl$ and $Ti(OC_4H_9)_4$ /17/. The authors showed that the effect of the organic groups on Ti and Al atoms on the yield of the cyclotrimers is insignificant. An important factor is the Al:Ti ratio (optimum value 20) and the number of halogen atoms in the organoaluminum component. If the value of n in $R_{3-n}AlX_n$ is more than 1.5 or less than 0.5, the polymerization becomes linear instead of cyclic: according to Wilke /18/, the system $(C_2H_5)_3Al — Ti(OR)_4$ catalyzes the formation of 1, 2-polybutadiene.

Müller /19/ cyclotrimerized butadiene with the aid of catalysts which, strictly speaking, did not contain organometallic compounds: a mixture of metallic aluminum, aluminum trichloride, titanium chlorides and NaCl in benzene. When $FeCl_3$ was substituted for the titanium compounds, a mixture of CTD isomers and linear oligomers was obtained /20/. Zakharkin proposed a method for the preparation of cis-trans-trans-cyclododecatriene-1, 5, 9 by cyclomerization of butadiene-1, 3 in a solution of toluene in the presence of the catalytic system of $TiAl_2Cl_8 \cdot C_6H_6$ and R_2AlR', where R' is alkyl or halogen at the molar component ratio 2:1.

It was shown by Wilke /21/ that in the presence of complex organometallic catalysts based on Ti and Cr, linear oligomers of isoprene and dimethylbutadiene can be formed. Subsequently, the butadiene dimer 3-methylheptatriene-1, 4, 6 was prepared in the presence of Co and Fe catalysts:

$$CH_2{=}CH{-}\underset{}{\overset{\overset{\displaystyle CH_2}{\|}}{C}H} + \overset{\overset{\displaystyle H}{|}}{C}H{=}CH{-}CH{=}CH_2 \longrightarrow CH_2{=}CH{-}\overset{\overset{\displaystyle CH_3}{|}}{C}H{-}CH{=}CH{-}CH{=}CH_2$$

The formation of this dimer was also noted by Otsuka et al. /22/ when butadiene was dimerized in the presence of $[Co(CO)_4]_2 — (C_2H_5)_3Al$. The other butadiene dimer — n-octatriene-1, 3, 6 — was formed as the side product.

A mixture of linear oligomers of butadiene may be obtained rapidly and in a high yield if cobalt catalysts are employed at 50 — 100°C /23/. These catalysts are formed by reducing cobalt salts in organic solvents in the presence of butadiene. The resulting brown-colored solutions probably contain butadiene complexes of cobalt, in which the valency of cobalt is zero. They react with butadiene to yield a mixture of dimers, trimers and tetramers of the general formula $(C_4H_6)_n$ with $(n + 1)$ double bonds, two of which are conjugated. The dimer fraction contains about 90% of 3-methylheptatriene-1, 4, 6, while the remainder consists of n-octatriene-1, 3, 6. If arylphosphates are added, the catalytic effect becomes radically changed: the main conversion products of butadiene in such a case are cycloocta-diene-1, 5 and vinylcyclohexene /24/. Similar results were also obtained by Japanese workers /25/. When cobalt was replaced by iron, the formation of a linear trimer was catalyzed /26/.

We may still mention another, recent method /27, 27a/ of stereospecific preparation of 1, 4-dienes from 1, 3-dienes and ethylene. The components of the catalyst system are triethylaluminum and an iron salt:

$$CH_2{=}\overset{\overset{\displaystyle R}{|}}{C}{-}\overset{\overset{\displaystyle R'}{|}}{C}{=}CHR'' + CH_2{=}CH_2 \longrightarrow {}^1/_2 CH_2{=}CH{-}CH_2{-}\overset{\overset{\displaystyle R}{|}}{C}{=}\overset{\overset{\displaystyle R'}{|}}{C}{-}CH_2R'' +$$

$$+{}^{1}/_{2}CH_3-\overset{R}{\overset{|}{C}}=\overset{R'}{\overset{|}{C}}-CHR''-CH=CH_2 \ (R, R', R''=H \text{ or } CH_3)$$

Cyclic trimerization of acetylenes

Acetylenic hydrocarbons and some of their derivatives, much like diolefins, may form linear polymers or ring trimers in the presence of an organometallic complex catalyst. Linear polyacetylenes form a conjugated system of double bonds and for this reason display a number of interesting properties. The preparation and properties of these polymers — potential organic semiconductor materials — is the subject of intense studies and numerous publications /28, 29, 30/. The preparation of both high-molecular polyacetylenes and of linear oligomers has been reported /31, 32/.

Of major interest is the use of complex catalysts in ring oligomerization of acetylenes to substituted benzenes. A review on the subject has been recently published /33/, and for this reason only the most interesting questions will be mentioned here.

During a few years following the appearance of the work of Ziegler on the utilization of complex polymerization catalysts it was believed that the "acid" hydrogen atom in acetylene is apt to destroy the catalyst. However, it was then shown by Natta et al. /28/ that both monosubstituted acetylenes and acetylene itself can become polymerized on such catalysts. The first communication concerning the preparation of cyclotrimers in the presence of Ziegler catalyst referred to disubstituted acetylenes /34/, but it was later found that cyclic trimers of monosubstituted acetylenes and of acetylene itself could also be obtained:

$$3RC{\equiv}CH \rightarrow \text{1,2,4-}R_3C_6H_3 \ + \ \text{1,3,5-}R_3C_6H_3$$

Franzus et al. /34/ were the first to effect cyclotrimerizations of a number of disubstituted acetylenes ($CH_3C\equiv CCH_3$, $C_2H_5C\equiv CC_2H_5$ and $C_6H_5C\equiv CC_6H_5$) under mild conditions in the presence of triisobutylaluminum — titanium tetrachloride catalyst. The yields of hexasubstituted benzenes were almost quantitative. None of the catalyst components produced cyclotrimerization by itself; when $TiCl_3$ was used instead of $TiCl_4$, the catalytic activity was lost.

A number of 1, 3, 5- and 1, 2, 4-trisubstituted benzenes were prepared from mono-substituted acetylenes by Lutz /35/ who used the system triethylaluminum — titanium tetrachloride. A more stereospecific system is triisobutylaluminum — titanium tetrachloride, which was used by Reikhsfel'd et al. /36/ to prepare 1, 3, 5-tributylbenzene (more than 90% yield) from hexyne-1 and 1, 3, 5-triphenylbenzene from phenylacetylene (the asymmetric

1, 2, 4-isomers were absent in the reaction products). These workers also showed /37/ that different acetylenic compounds could be co-cyclotrimerized. In this way it proved possible to prepare cyclic trimers of acetylenes which do not undergo cyclotrimerization when taken separately /38/. A study of this reaction in the presence of various titanium-based catalyst systems /39/ confirmed that the active site of the reaction is the titanium atom. Cyclotrimerizations of certain acetylenes to alkylbenzenes have been realized in the presence of complex catalysts; such reactions are very difficult to realize by any other method. Thus, both isomers of tri-tert-butylbenzene (1, 3, 5- and 1, 2, 4-isomers) were first prepared directly when the system triisobutylaluminum — titanium tetrachloride was used as a catalyst /40/. Owing to the considerable steric hindrances offered by the bulky tert-butyl groups, previous attempts at cyclotrimerization of tert-butylacetylene on other catalysts had been unsuccessful.

It was shown by Hoover et al. /41/ that vinylacetylene can be converted to trivinylbenzene in the presence of Ziegler catalyst at 0 — 10°C. It is seen that triple bonds only participate in this reaction.

The formation of benzene derivatives from acetylenes is not specific to titanium catalysts. Titanium tetrachloride may be replaced by VCl_4 /42/, $VOCl_3$ /42, 43/, CrO_2Cl_2 /44/, or $CoCl_2$ /45/. Nevertheless, systems containing tetravalent titanium are the most active (in the presence of trivalent titanium the polymerization becomes linear /29, 36/) if the ratio $AlR_3:TiCl_4 = 3$ to 4. Cyclic trimerization of isopropenylacetylene in the presence of different catalysts has been recently studied /46/; Ziegler catalysts gave the highest yields of the cyclic products.

A catalyst system other than complex organometallic compounds was proposed by Luttinger /47/ for the cyclization of acetylenes. It consists of a salt of a Group VIII metal and a hydride reducing agent (alkali metal borohydrides and aluminohydrides). The highest yields of benzene derivatives were obtained with cobalt salts, irrespective of the valency of this metal in the initial salt. The system $NiCl_2 — NaBH_4$ proved the most suitable for linear polymerization of acetylenes.

2. ISOMERIZATION OF OLEFINS IN THE PRESENCE OF COMPLEX CATALYSTS AND THEIR COMPONENTS

Organometallic complex catalyst systems may produce isomerization of olefins. Chauvin and Lefebre /48/ studied the rate of migration of the double bond in different substituted olefins ($C_4 — C_6$) in the presence of catalysts formed by the reaction between triethylaluminum and salts of transition metals. They showed that the catalytic effect is significant only if the system contains chromium, iron or nickel salts. Thus, pure butene-1 was passed through a solution of the catalyst $(C_2H_5)_3Al — NiCl_2 \cdot 2C_5H_5N$ (molar ratio $Al:Ni = 4:1$) in pentane or hexane at the boiling temperature of the solvent. The result was evolution of gas consisting of 71.9% of trans-butene-2, 25.6% of cis-butene-2 and 2.5% of the unreacted butene-1. Back isomerization of trans-butene-2 did not take place under these conditions; insignificant conversion of this compound to the cis-isomer and butene-1 was noted. The authors studied the isomerization of a number of pentenes and hexenes. They showed that the composition of the isomerization products of butene-1 is close to that thermodynamically calculated, but this was not the case for the isomerization of pentenes. 4-Methylpentene-1 and 4-methylpentene-2 displayed a very weak migration of the double bond to the tertiary carbon atom. The activity of the isomerization catalyst sharply decreases when triphenylphosphine is added to the system. In the view of the authors, olefin isomerization proceeds via an intermediate π-allyl complex.

A patent has been taken out /49/ for the isomerization of olefins in the presence of a system constituted by a trialkylaluminum compound and a tetraalkyl o-titanate (or vanadate), at a molar ratio of aluminum to titanium between 2 and 8. These systems catalyze, first and foremost, the isomerization of unbranched olefins. Thus, butene-1 yielded a mixture of cis-butene-2 (67%) and trans-butene-2 (28%) within 15 minutes at 20°C.

Isomerization of olefins during polymerization effected on Ziegler-Natta catalysts is the subject of a number of recent studies /50 — 56/.

It was found by Symcox /50/ that the polymerization of butene-2 and pentene-2 in the presence of $(C_2H_5)_3Al - TiCl_4$ catalyst between −10°C and +100°C yields polybutylene-1 and polypentene-1. Formation of high-molecular, isotactic polybutene-1 was noted during the polymerization of cis- and trans-butene-2 in the presence of the system $(C_2H_5)_3Al - TiCl_3$ /51/. These workers found that, in the presence of a catalyst, butene-2 is isomerized to an equilibrium mixture of butene-1, cis- and trans-butene-2; the composition of the mixture is in agreement with that calculated from the free energies of formation. Thus, butene-2 is first isomerized to butene-1, which then polymerizes in the conventional manner.

A comparative study of the polymerization of a number of α- and β-olefins in the presence of $AlR_3 - TiCl_3$ catalysts at 80°C showed /52/ that α-olefins polymerize practically without isomerization (polymerization rate higher than isomerization rate), while all unbranched β-olefins give an isomer distribution which is close to the equilibrium value. Branched β-olefins do not seem to be isomerized to α-olefins at all.

Aubrey et al. /53/ recently showed by IR spectroscopic and chromatographic methods that when octene-1 or octadecene-1 is polymerized in the presence of $Al(C_2H_5)_3 - TiCl_4$ catalyst, the nonpolymerized olefin contains trans-1, 2-substituted ethylene (the cis- isomer is not formed), the amount of which increases with the concentration of the catalyst, temperature and duration of the reaction, and with the change in the Al:Ti ratio (pure $(C_2H_5)_3Al$ does not produce isomerization). The isomer with nonterminal double bonds does not copolymerize with the α-olefin, but is not inert during the polymerization. If its concentration in the starting olefin is high, the yield and the molecular weight of the resulting polymer decrease.

Schindler /54/, who studied the isomerization of olefins which occurs during their polymerization on Ziegler-Natta catalysts, using deuterated ethylenes, assumes that the catalyst has two active sites, one of which is responsible for the chain growth, while the other is responsible for the isomerization. Isomerization takes place only in the presence of catalysts which contain active sites, preferably those containing Ti^{2+} ions. Such catalysts contain a strong reducing agent (AlR_3), while their Al:Ti ratio is large. If AlR_3 is replaced by AlR_2Cl, isomerization cannot take place. According to Schindler, the isomerization proceeds as follows:

$$\text{cat}-H+\underset{CH_3}{\overset{CH_3}{|}}CH{=}CH \rightleftarrows \text{cat}-\overset{CH_3}{\underset{CH_2-CH_3}{CH}} \rightleftarrows \text{cat}-H+\overset{CH_3}{CH}{=}\underset{CH_3}{CH}$$

$$\downarrow$$

$$\text{cat}-H+CH_2{=}CH-CH_2-CH_3 \longrightarrow \text{polymer}$$

(cat-catalyst complex) c)

The presence of individual components of Ziegler catalysts and their conversion products may bring about secondary changes in the polymeric polydiene chain (cis-trans-isomerization, migration of double bonds, cyclization). This is the result of the high cationic activity of the compounds employed as constituents of Ziegler catalysts (R_2AlCl, $RAlCl_2$, $AlCl_3$, $TiCl_4$, etc.). It is known that the internal double bonds are very reactive in cationic processes.

If solutions of natural rubber are treated with reagents such as AlR_2Cl, $AlRCl_2$ or $TiCl_4$ (60°C, 12 hours), the number of cis-1, 4-links in the chain decreases considerably, and the total unsaturation decreases somewhat as well /57/. During cationic polymerization of dienes the growing polymer cation readily reacts with the double bond of the chain proper; this in conjugation with favorable steric probability results in the formation of six-membered ring structures. Similar considerations apply to the reaction between the polymer chain and cation-active substances /57 — 61/. It was shown by Ermakova /57/ that such reactions are intramolecular even in concentrated solutions:

$$\sim CH_2CH{=}CHCH_2CH_2CH{=}CHCH_2CH_2CH{=}CHCH_2\sim$$

$$\downarrow R^+$$

$$\sim CH_2CH{=}CHCH_2CH_2CH{=}CHCH_2CH_2CH\overset{\oplus}{C}HCH_2\sim \quad (\text{with } R \text{ on } CH)$$

$$\downarrow$$

$$\sim CH_2CH{=}CHCH_2CH_2-CH$$

$$\sim -CH_2-CH\ \overset{\oplus}{C}H \quad | \quad R-CH\ \ CH_2 \quad \backslash CH_2 / \longrightarrow \sim-[\text{bicyclic ring}]-\sim$$

Cationic primers (e.g., alkylaluminum halides activated by water, etc.) may produce migration of hydrogen along the chain with formation of conjugated double bonds /57/:

$$-CH_2CH{=}CH-CH_2CH_2CH{=}CHCH_2\sim$$

$$\downarrow$$

$$-CH_2CH{=}CH-CH_2CH{=}CH-CH_2-CH_2\sim$$

$$\downarrow$$

$$-CH_2CH{=}CH-CH{=}CH-CH_2CH_2CH_2\sim$$

The high reactivity of internal double bonds in cationic processes is the reason for their ready cis-trans-isomerization in the presence of alkylaluminum halides, aluminum trichloride, titanium tetrachloride, etc. When cation-active components add on to the double bond, the cyclization is preceded by the formation of a π-complex, which seems to readily undergo cis-trans-transitions /57 — 58/.

Ermakova /57/ studied the isomerization of 3-methylpentene-2 as a model of the repeating unit of polyisoprene chain, in the presence of $C_2H_5AlCl_2$ and $TiCl_4$. After this olefin had been heated at 60°C for 3 hours with $C_2H_5AlCl_2$ (3.6 mole %) the concentration of the cis-isomer decreased from 85.6 to 43.8%, while the concentration of the trans-isomer increased correspondingly.

We must also mention the very interesting studies on the 1, 3-polymerization of olefins in the presence of complex catalysts. Yuguchi and Iwamoto /62/ studied the polymerization of propylene with a catalyst produced by the reaction between triethylaluminum with the adduct of VCl_4 and ferric acetylacetonate (1:1). It was found that, depending on the experimental conditions (in the first place on the identity of the solvent employed), the product consists either of polypropylene of conventional structure or of ethylene-propylene copolymer and even of pure polyethylene. The latter structures may be formed if a hydrogen atom migrates from the methyl group to the neighboring carbon atom and the molecule is then polymerized in position 1, 3:

$$CH_2{=}CH(CH_3) \rightarrow \begin{cases} \left[-CH_2-CH(CH_3)-\right]_x \\ \left[-CH_2CH_2CH_2-\right]_y \end{cases}$$

The degree of displacement of the hydrogen atom during polymerization will depend on the identity of the solvent: ordinary polypropylene was obtained in heptane; ethylene-propylene copolymer was obtained in benzene (especially at elevated temperatures and pressures); crystalline polyethylene was obtained in dichloroethane under atmospheric pressure. Infrared spectra of the polymer formed in benzene show a weak absorption band at 1156 cm^{-1}, which indicates that the methyl group content in the polymer is low. The polymer is an amorphous copolymer of ethylene and propylene, which contains up to 58% of ethylene links. X-ray study of the polymer obtained in dichloroethane confirmed that its structure is identical with that of crystalline polyethylene.

It was shown by Gol'dfarb at an earlier date /63/ that isobutylene can be 1, 3-polymerized. The polymer was prepared in the presence of triethylaluminum — $TiCl_4$ catalyst, at an Al:Ti ratio above 5 and its spectra showed a very high content of methylene groups. This means that the polyisobutylene product contained not only repeating units of conventional structure, but also polypropylene links:

$$CH_2{=}C(CH_3)_2 \rightarrow \begin{cases} \left[-CH_2-C(CH_3)_2-\right]_x \\ \left[-CH_2-CH(CH_3)-CH_2-\right]_y \end{cases}$$

Korneeva et al. /64/ recently studied the polymerization of isobutylene in the presence of the catalyst systems $(C_2H_5)_3Al-VOCl_3$, $C_2H_5Al[N(SiR_3)_2]_2-VOCl_3$ and $C_2H_5Al[N(SiR_3)_2]_2-TiCl_4$. These workers showed that the polymerization takes place on the complex as a whole and not on its individual components. Study of IR spectra of the polymer product showed that its methylene group content was much higher than that of ordinary polyisobutylene, as indicated by the changes in the optical density ratio of the 2925 cm^{-1} (symmetrical C — H stretching vibrations in CH_2-groups) to the 2960 cm^{-1} band (symmetrical C — H stretching vibrations in CH_3-groups). This ratio was 0.64 in polymer prepared in the presence of $TiCl_4$, but was 0.9 and 1.13, respectively, in polymers prepared in the presence of $(C_2H_5)_3Al-VOCl_3$ and $C_2H_5Al[N(SiR_3)_2]_2-VOCl_3$.

3. ALKYLATION CATALYSTS

Since aluminum chloride is one of the conventional catalysts of alkylation of aromatic compounds, it could be expected that alkylaluminum halides, too, would catalyze alkylation reactions. In fact, it was shown in 1957 by Groizeleau /65/ that a small amount of hexaethylbenzene is formed by the action of ethylaluminum sesquibromide on benzene. The reaction between benzene and ethyl bromide in the presence of $(C_2H_5)_3Al_2Br_3$ yielded up to 40% hexaethylbenzene, while not exceeding 1.5% when aluminum chloride or aluminum bromide was used as catalyst. This catalyst was also employed by Nicolescu et al. /66/ in the alkylation of aromatic hydrocarbons by cyclohexene. When cyclohexene was made to react with benzene in the presence of $(C_2H_5)_3Al_2Br_3$ at 80 — 120°C and under atmospheric pressure (benzene : cyclohexene = 1 : 1.5), the yield of the alkylate attained 76.5%. The alkylate contained up to 20% monocyclohexylbenzene and more than 30% dicyclohexylbenzene. The authors pointed out that aluminum catalyst in the form of an organic compound is much to be preferred to $AlCl_3$: it is much more readily soluble in aromatic compounds and — most important — is much more active than $AlCl_3$. A high yield of the alkylate is obtained even if the catalyst concentration is as low as 0.01 mole of the catalyst per 1 mole of cyclohexene (for $AlCl_3$, 0.12 — 0.3 moles of the catalyst are needed per mole of the olefin). It is assumed that the alkylation involves the formation of an intermediate triple complex of the organoaluminum compound, aromatic hydrocarbon and olefin:

$$C_6H_6 + R_2AlX \rightleftarrows [C_6H_5AlR_2X]H \xrightarrow{\;>C=C<\;} [C_6H_5AlR_2X]\overset{|}{\underset{|}{C}}-\overset{|}{\underset{|}{C}}H$$

This complex is then decomposed with the formation of an organoaluminum compound:

$$[C_6H_5AlR_2X]\overset{|}{\underset{|}{C}}-\overset{|}{\underset{|}{C}}H- \rightleftarrows C_6H_5-\overset{|}{\underset{|}{C}}-\overset{|}{\underset{|}{C}}H + R_2AlX$$

A patent /67/ has been taken out for the alkylation of benzene by olefins with three carbon atoms or more in the presence of $AlRCl_2$ as catalyst. Benzene smoothly reacts with propylene in this way even at 25°C. It is interesting to note that propylene alone is the reactive species in the mixture $C_3H_6 - C_2H_4$. The reaction product contained 48.9% of unreacted benzene, 34.9% of isopropylbenzene, 12.2% of diisopropylbenzenes and 4% of triisopropylbenzenes. Ethylbenzene was not detected.

According to another patent /68/ alkylbenzenes with 8 — 16 carbon atoms in the side chain are prepared in the presence of systems much like complex organometallic catalysts. It is interesting to note that the process includes an alkylation and a propagation reaction. The catalyst system consists of AlX_3, AlR_nX_{3-n} (R is alkyl, aryl or cycloalkyl) and halides of Ti, V, Zr, Cr or Fe. If it is desired to obtain the maximum yield of the most important fraction (b. p. up to 216°C / 1 mm, i. e., C_8H_{17} to $C_{16}H_{33}$ in the side chain), the ratio of AlX_3 to AlR_nX_{3-n} should be 0.3 to 0.1, while the ratio between the number of alkyl groups in AlR_nX_{3-n} and the total number of moles of the transition metal halide should be 0.1 to 1.0. Thus, when 3 liters of dry benzene, containing 10 g $AlCl_3$, 1.4 ml of 92% $(C_3H_5)_3Al$ and 1.9 g $TiCl_3$, were reacted with ethylene at 30°C, 324 grams of the desired fraction (38% of total alkylate) were obtained, while at 50°C, 398 grams (61% of total alkylate) were obtained.

Ethylation of benzene in the presence of catalysts based on alkylaluminum chlorides and alkyl chlorides was studied by Nicolescu and Serban /69/.

4. UTILIZATION OF COMPLEX CATALYSTS AS REDUCING AGENTS

Catalysis of hydrogenation

Complex metal hydrides are widely used in chemical practice as hydrogenation agents /70/. It was recently found that the reaction between these compounds and salts of transition metals yields highly effective catalysts of hydrogenation of olefins. Takegami et al. /71/ used systems based on $LiAlH_4$ and $CoCl_2$ or $FeCl_3$ and carried out /72/ a detailed investigation of the hydrogenation of styrene in the presence of the catalyst system $LiAlH_4 - FeCl_3$. An active catalyst is obtained if the ratio $FeCl_3$: : $LiAlH_4$ is unity. The solvents which were employed included several ethers (diethyl ether, dibutyl ether), tetrahydrofuran, dioxane and diethylene glycol. It was shown that the rate of hydrogenation of styrene in various solvents varies in the following sequence:

dibutyl ether > diethyl ether > tetrahydrofuran > diethylene glycol > dioxan

The active catalyst is not formed in hydrocarbon solvents, but the addition of a small amount of tetrahydrofuran results in the appearance of catalytic activity. More basic compounds (pyridine, triphenylphosphine, etc.) substantially reduce the activity of the system.

Various organoaluminum compounds can be used instead of the complex metal hydride to catalyze the hydrogenation of olefins. Heptane-soluble and toluene-soluble catalysts have been reported /73/ which are active under mild conditions. They have AlR_3 and tetraalkoxytitanium as their constituent compounds and are active at 25 — 50°C under 2.4 — 2.7 atm hydrogen pressure. A complex hydrogenation catalyst, formed by the reaction between trialkylaluminum and a transition metal carboxylate has been proposed (e. g., nickel 2-ethylhexanoate and triethylaluminum) /74/.

A very detailed study of olefin hydrogenation in the presence of Ziegler-type homogeneous catalysts was carried out by Sloan et al. /75/. Catalysts were obtained by the reaction between triethylaluminum or triisobutylaluminum and compounds of Ti, Zr, V, Cr, Mo, Mn, Fe, Co, Ni, Pd and Ru, mostly acetylacetonates, and in the case of Ti, V and Zr also with alkoxides. All these substances catalyzed to some extent the hydrogenation of various mono-, di-, tri- and tetrasubstituted olefins (octene-1, cyclohexene, etc.). The hydrogentation was carried out under mild conditions (30 — 50°C, about 3.5 atm hydrogen pressure). The most active systems were those based on triisobutylaluminum and iron and cobalt acetylacetonates. The latter system (Al:Co = 6:1) ensured a complete hydrogenation of cyclohexene within only 20 minutes at 30°C. In addition, the system $(C_5H_5)_2TiCl_2 - Al(C_2H_5)_3$ (equimolar ratio), which is homogeneous during the hydrogenation, was also studied; octane was obtained in 70% yield from octene-1 at room temperature. It was shown that if butyllithium is employed instead of an alkylaluminum compound, the hydrogenation rate decreases markedly. According to the authors, the true catalyst species is the hydride compound of the transition metal, which is formed as follows:

$$AlR_3 + MX_n \longrightarrow R_2AlX + RMX_{n-1}$$

where M is the transition metal and X are the groups bound to it.

The organometallic compound RMX_{n-1} reacts with hydrogen, with formation of hydride

$$RMX_{n-1} + H_2 \longrightarrow RH + HMX_{n-1}$$

which adds onto the double bond of the olefin:

$$>C=C< + HMX_{n-1} \longrightarrow H-\overset{|}{\underset{|}{C}}-\overset{|}{\underset{|}{C}}-MX_{n-1}$$

The newly formed organometallic compound reacts with hydrogen, when HMX_{n-1} is again liberated:

$$H-\overset{|}{\underset{|}{C}}-\overset{|}{\underset{|}{C}}-MX_{n-1} + H_2 \longrightarrow H-\overset{|}{\underset{|}{C}}-\overset{|}{\underset{|}{C}}-H + HMX_{n-1}, \text{ etc.}$$

A recent publication /76/ deals with the results of a kinetic study of the hydrogenation of olefin (cyclohexene and heptene-1) in the presence of homogeneous catalysts prepared from triethylaluminum and $(C_5H_5)_2TiCl_2$ or V, Cr, Mn, Fe, Co and Ni acetylacetonates. It was found that the stage which determines the reaction rate is the interaction of the initially formed transition metal hydride with the olefin. At the ratio Al:M = 10, the activity of the catalysts of hydrogenation of cyclohexene decreases in the sequence Co > Ni > Fe > Cr. A similar series is also obtained for the hydrogenation of heptene-1:

$$CO > Ni > Fe > Cr > Ti > Mn > V$$

The hydrogenation reaction is then complicated by isomerization. One or two minutes after the catalyst and the olefins have been mixed together, heptene-2 and heptene-3 could be chromatographically detected in the sample.

After the completion of the hydrogenation heptane alone was present in the sample, which shows that the isomerization merely consisted in a shift of the double bond. Hydrogenation of alkylaromatic hydrocarbons have also been studied /76a/.

It will be noted that the employment of the above hydrogenation catalysts is of practical interest, because the reaction can be conducted under mild conditions; since the system is homogeneous, each metal atom is able to participate in the catalysis.

Preparation of metal carbonyls (reductive carbonylation)

The reducing properties of alkylaluminum compounds were utilized by several workers to prepare carbonyl complexes of certain transition metals. Zakharkin et al. /77/ prepared carbonyls of Cr, Mo and W by reacting triethylaluminum or diisobutylaluminum hydride with the metal chloride in the presence of carbon monoxide. The reaction was conducted in an ether solution at 120 — 130°C and under 30 — 50 atm pressure of carbon monoxide. When triethylaluminum was employed, the yields of the final products were as follows: $Cr(CO)_6$ 54%; $Mo(CO)_6$ 70%; $W(CO)_6$ 86%; they were somewhat higher when diisobutylaluminum hydride was used instead of triethylaluminum.

Podall /78/ used the same method of preparation of these rare compounds. He noted that the yields of metal carbonyls prepared in the presence of alkylaluminum compounds were much higher than those obtained by Grignard synthesis, and that almost no free metal was formed. The synthesis of metal carbonyls was conducted in two steps. Thus, for instance, a suspension of $CrCl_3$ in ether was mixed with an ethereal solution of triethylaluminum at 0°C, after which carbon monoxide was fed in and the mixture was heated at 100 — 115°C under about 200 atm pressure for $5\,^1/_2$ hours. The yield of $Cr(CO)_6$ was 76% under these conditions. Subsequently /79/ this worker reported the synthesis of manganese carbonyl, which is very difficult to prepare. Podall et al. /80/ made a special study of the effect of different factors on the syntheses of carbonyls of Cr, Mo, W and Mn. They showed that the yield of carbonyl strongly depends on the nature of the initial metal salt. In the system metal salt — triethykaluminum in benzene the yields of the carbonyls decreased in the following sequence:

$$WCl_6 > MoCl_5 > Cr(C_5H_7OO)_3 > Mn(OCOCH_3)_2 > CrCl_3 \gg MnCl_2$$

The carbonylation process is also materially affected by the nature of the solvent. For the system $CrCl_3 - (C_2H_5)_3Al$ (Al:Cr = 6:1) at 115°C, the yield of $Cr(CO)_6$ in diethyl ether was 85%, in benzene 20%, and in pyridine only 10%. The best solvent for the preparation of manganese carbonyl ($Al(C_2H_5)_3 : Mn(OCOCH_3)_2$ = 4:1) was diisopropyl ether (yield 56%), while the yield in pyridine was less than 1%. The yield of manganese carbonyl may be increased to 80% if the Al:Mn ratio is increased to 9:1, but this is technically inconvenient in preparative work. Table 16 shows the optimum conditions for the syntheses of chromium, molybdenum and tungsten carbonyls under ~ 200 atm pressure of carbon monoxide (yields below 90%). The activity of the system markedly depends on the identity of the organoaluminum compound. The synthesis of manganese carbonyl from manganous acetate in benzene at 100°C and Al:Mn = 4:1 showed that the activities of the organoaluminum components decreased in the following sequence:

$$(iso\text{-}C_4H_9)_3Al > (C_2H_5)_3Al > (C_2H_5)_2AlH > (CH_3)_3Al > (C_2H_5)_3Al_2Cl_3$$

The yields of the carbonyl product decrease from 45% to 1% in this series. On the contrary, trimethylaluminum is more active than triethylaluminum or triisobutylaluminum in the preparation of chromium, molybdenum and tungsten carbonyls.

TABLE 16. Synthesis of metal carbonyls

Initial salt	Ratio AlR_3 : salt	Temperature, °C	Solvent
$CrCl_3$	5:1	115	Ether, benzene
$MoCl_5$	4:1	65	Benzene
WCl_6	3:1	50	"

It was noted that if much time is allowed to elapse between the preparation of the catalyst and the delivery of carbon monoxide, the metal may separate out as element and the carbonylation reaction is then much slower. It is believed that triethylaluminum reacts with the metal salt to form ethyl derivatives of the metals, which may either decompose with the liberation of elementary metal, or may react with CO to yield the metal carbonyl. The alkylation of the metal is the rate-determining stage.

The process of "reductive carbonylation" in its different varieties forms the subject of several US patents. The use of organic compounds of B, Ga, Zn, Cd and other metals is recommended /81/ and also the use of Group Ia metals /82, 83/. A method for the preparation of chromium, molybdenum and tungsten carbonyls by reaction with triethylaluminum in tetrahydrofuran under low pressures of carbon monoxide has been patented /84/.

Szabó and Markó /85/ recently reported the preparation of cobalt carbonyl by this method /79, 80/. The catalyst was prepared by mixing hexane solutions of cobalt stearate with triethylaluminum. About 1.5 mole of CO per mole of cobalt was absorbed during 16 hours at room temperature. The highest rates of absorption were noted when the Al:Co ratio in the catalyst was between 1 and 6. The main product was dicobaltooctacarbonyl $Co_2(CO)_8$; $Co(CO)_{12}$ and propionylcobalt tetracarbonyl $C_2H_5COCo(CO)_4$, identified as the adduct with triphenylphosphine, were also isolated. The last-named compound is formed by the attack on the Co — C bond by the carbon monoxide molecule and proves that an alkylcobalt compound is formed in the first stage of the process.

Preparation of π-complexes of transition metals

The review by Bennett /86/ on the subject of olefin and acetylene complexes of transition metals deals with the general problem of preparation of metal complexes by treating the metal salt with an olefin in the presence of a strong reducing agent. The author noted that the method is not frequently used, since the reducing agent itself often decomposes the olefin or the resulting complex. However, as early as 1960 Wilke /44/ succeeded in preparing a nickel complex of cyclododecatriene by reacting nickel acetylacetonate with trans-cyclododecatriene in the presence of an organoaluminum compound.

The method in fact proved to be of general application, including the preparation of several unstable complexes. According to Wilke /16/, the principle of the method is the following: transition metal compounds in the presence of electron donors (olefins or acetylenes as well as phosphines, arsines, etc.) are reduced under suitable conditions by metal alkyls or metal hydrides. Complexes are formed, in which the transition metal has a lower valency than in the starting compound. According to Wilke, the reaction involves the formation of unstable alkyls of transition metals as the intermediate stage. The preparation of many complexes of transition metals by this method has been patented /87, 88/: the complexes include $[C_{12}H_{18}]Ni$; $[C_8H_{16}]_2Ni$; $Ni[P(C_6H_5)_3]_4$; $Ni[P(C_6H_5)_3]_2$; $Co[P(C_6H_5)_3]_4$; $(C_8H_{17})_{1-2}Co$; $(C_8H_{14})_4Ti$; $Pd[P(C_6H_5)_3]_4$; $Ni[Sb(C_6H_5)_3]_4$; $Ni[As(C_6H_5)_3]$. Many complexes proved to be active in the oligomerization of diolefins /89/. The reducing agents used in the preparation of these complexes included trimethylaluminum, triethylaluminum and, most often, $(C_2H_5)_2AlOC_2H_5$.

Wilke and Herrmann /90/ prepared ethylene complexes of the type $(R_3P)_2Ni \cdot C_2H_4$, where R is phenyl, ethyl, cyclohexyl, etc., in a similar manner. These compounds were obtained in almost quantitative yield by the reaction between nickel acetylacetonate and PR_3 in the presence of $(C_2H_5)_2AlOC_2H_5$ (Ni : P = 1 : 2) in a solution of benzene saturated with ethylene. Complexes containing styrene, α-methylstyrene, etc., were prepared by displacement of ethylene from these compounds.

Tsutsui and Chang /91/ employed this general method to obtain π-complex bis-arene complexes of transition metals. They heated mixtures of the transition metal halide, triethylaluminum and aromatic hydrocarbon in a normal aliphatic solvent (heptane) for 10 — 30 minutes at 130 — 140°C. After cooling to below 0°C methanol was added and the resulting mixture was hydrolyzed by water in the air. In most cases the hydrolyzate contained the π-complex ion as hydroxyl, which could be isolated as salt. Complexes of chromium with toluene, xylene, mesitylene, diphenyl and fluorene, as well as bis-mesityleneiron were prepared in this way. Bis-arene complexes of chromium, with benzene rings interconnected by a polymethylene chain have also been prepared /92/; thus, for instance, 1, 1-(β, γ-diphenyltetramethylene)-bis-benzenechromium (compound I) was prepared by reacting $CrCl_3$ with trans-stilbene and triethylaluminum in boiling heptane. It was precipitated as iodide and was recrystallized from chloroform and heptane:

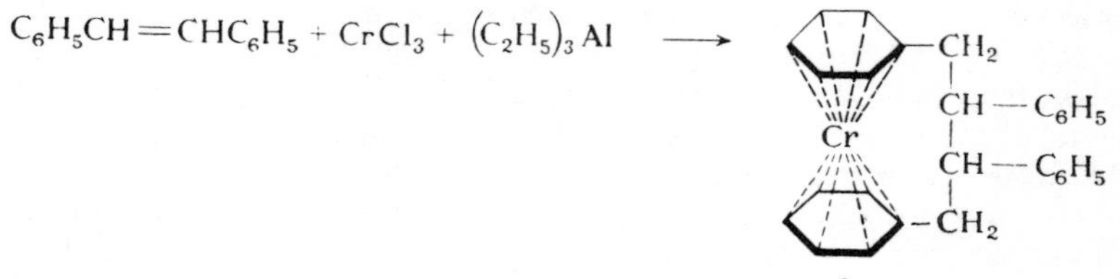

$$C_6H_5CH{=}CHC_6H_5 + CrCl_3 + (C_2H_5)_3Al \longrightarrow$$

Cr; CH_2; $CH-C_6H_5$; $CH-C_6H_5$; CH_2

I

According to Tsutsui and Chang /91/, the reduction of the transition metal compound results in the formation of an organometallic compound, which decomposes to yield organic radicals and "radical" metal, which is not in its ground state and which yields the bis-arene complex in the presence of aromatic donor molecules.

It should be pointed out that the intermediate organometallic compound can be stabilized under suitable conditions, e.g., if electron-donating ligands are present in the complex. Thus, dipyridyl complexes of iron and nickel with σ-bonds between the metal and the ethyl group could be prepared /93/. They were prepared by the reduction of iron and nickel acetylacetonates by $(C_2H_5)_2AlOC_2H_5$ or $(C_2H_5)_3Al$ in ether in the presence of α, α'-dipyridyl between $-20°C$ and $0°C$; the yields were 50% Fe and 70% Ni. The complexes are stable out of contact with air:

$$(\alpha,\alpha'\text{-dipy})_2Fe(C_2H_5)_2 \qquad (\alpha,\alpha'\text{-dipy})Ni(C_2H_5)_2$$

Both these complexes actively catalyze the cyclooligomerization of butadiene.

Dialkylnickel compounds R_2Ni, stabilized with α, α'-dipyridyl ($R = CH_3$, C_2H_5) or with two molecules of tris-(2-biphenylyl) phosphite ($R = CH_3$), were recently prepared by Wilke and Herrmann /94/. The last-named complex was prepared by the reaction:

$$Ni(C_5H_7O_2)_2 + 2(CH_3)_3Al + 2P(OC_6H_4C_6H_5)_3 \xrightarrow[\text{toluene}]{-45°C}$$

$$\longrightarrow [P(OC_6H_4C_6H_5)_3]_2Ni(CH_3)_2 + 2(CH_3)_2Al(C_5H_7O_2)$$

We may also mention the preparation of a very unusual substance — bis-tritylnickel — which is unusually stable for an alkyl of a transition metal /95/. This compound was prepared in over 90% yield by reacting a solution of nickel acetylacetonate in benzene with $(C_2H_5)_2AlOC_2H_5$ in the presence of excess hexaphenylethane at 0°C:

$$Ni(C_5H_7O_2)_2 + 2(C_2H_5)_2AlOC_2H_5 + (C_6H_5)_6Cr \longrightarrow$$

$$\longrightarrow (C_6H_5)_3CNiC(C_6H_5)_3 + 2(C_2H_5)_2Al(OC_2H_5)(C_5H_7O_2)$$

Chemical fixation of molecular nitrogen

Complex organometallic catalysts have recently found a new, very promising field of application: they are used in fixation of atmospheric nitrogen. The problem of converting molecular nitrogen to chemical compounds which would be assimilable by living organisms is of paramount scientific and economic importance. The methods employed at present (synthesis of ammonia etc.) require very drastic operating conditions: high temperatures, high pressures, etc. These processes are, as a rule, catalytic, but the catalysts employed are greatly inferior to the biochemical catalysts produced by different microorganisms (certain algae, tuber bacteria, etc.). These organisms assimilate nitrogen under mild conditions and at a rate which is beyond the wildest dreams of present-day technology.

The nature of the active sites responsible for the activation of the nitrogen molecule is still unclear, but a very important point which must be borne in mind is that the biological nitrogen-fixing systems all contain compounds of a number of transition metals (Mo, Fe, Co, etc.). It is believed /96/ that the nitrogen molecule interacts with the metal-bearing enzyme as follows:

$$\Phi - Me + N_2 \longrightarrow \Phi - Me \leftarrow \begin{matrix} N \\ ||| \\ N \end{matrix} \xrightarrow{[H]} NH_3$$

Here, $\Phi - Me$ is an enzyme which contains at least one atom of the transition element. Other enzymes supply active hydrogen, as a result of which the bound nitrogen is converted to a readily assimilable form.

It could accordingly be expected that π-complexes of N_2 with compounds of transition metals would be obtainable outside a biological system as well.

Catalytic systems which produce polymerization of olefins and acetylenes were considered first /96/. It had been noted, in fact /96, 97/, that catalyst systems constituted by a salt of a transition metal and an organic compound of Li, Mg, Al (or $LiAlH_4$) are capable of fixing molecular nitrogen. The nitrogen fixation was directly confirmed by experiments involving the use of radioactive isotope ^{15}N /98/.

TABLE 17. Absorption of molecular nitrogen by various catalyst systems (10 hours at room temperature under 150 atm pressure) /96/

Catalyst system	Solvent	Yield of NH_3 during hydrolysis, moles per 1 mole of transition metal
$CrCl_3 - LiAlH_4$*	Ether	0.02
	"	0.07
$CrCl_3 - C_2H_5MgBr$	"	0.017
$MoCl_5 - C_2H_5MgBr$	"	0.08
$WCl_6 - C_2H_5MgBr$	"	0.15
$FeCl_3 - C_2H_5MgBr$	"	0.09
$TiCl_4 - C_2H_5MgBr$	"	0.10
$TiCl_4 - (iso\text{-}C_4H_9)_3Al$	Heptane	0.25
$(C_5H_5)_2TiCl_2 - C_2H_5MgBr$*	Ether	0.67
	"	0.84
	Tetrahydrofuran	0.40
$(C_5H_5)_2TiCl_2 - C_4H_9Li$	Heptane	0.50
$CrCl_3 - CH_3MgI$	Ether	0.04
$CrCl_3 - C_2H_5MgBr$	"	0.17
$CrCl_3 - C_3H_7MgBr$	"	0.30

* Pressure 1 atm.

Table 17 shows the results of Vol'pin and Shur /96/ on the chemical fixation of nitrogen on various systems consisting of a salt of a transition metal and an organic compound of a Group I — III metal. It is seen from these data that the amount of the reacted nitrogen depends on several factors: nature of solvent (absorption decreases with increasing solvation capacity of the solvent), nature of the organic group bound to the metal, the pressure in the system, etc. Another important factor is the ratio between the organometallic compound and the transition metal salt (optimum values: 9 moles $LiAlH_4$ or RMgX and 3 moles of AlR_3 per 1 mole of the transition metal). The absorption of nitrogen was fully suppressed when substances capable of competing with nitrogen in complex formation (carbon monoxide, diphenylacetylene) were introduced as additive to the reaction mixture. This is also the reason for the low yields of ammonia when salts of transition metals are replaced by their phosphine complexes: $[P(C_6H_5)_3]_2TiCl_4$, $[P(C_6H_5)_3]_2FeCl_3$, etc. /99/. When molecular hydrogen was introduced together with nitrogen into the catalyst systems, the activity of the system remained unaffected or else gradually decreased.

It was subsequently noted, however /100/, that when homogeneous hydrogenation catalysts were employed, the absorption of nitrogen by the system $Ti(OC_2H_5)_4 + (iso\text{-}C_4H_9)_3Al$ (Al : Ti = 6) in toluene considerably increased on the introduction of molecular hydrogen. The absorption of nitrogen may also take place from nitrogen-hydrogen mixture, if the active catalyst complex is first deposited on a carrier /101/. This was experimentally demonstrated for the system $FeCl_3 - C_4H_9Li$ (1 : 10) on alumina and other solids as carriers.

Fixation of molecular nitrogen by systems of salts or complexes of transition metals, magnesium and iodine (molar ratio $MeX : Mg : I_2 = 1 : 10 : 5$) was recently described /102/. Such systems are often more active than those containing organometallic compounds. Highest yields of NH_3 were obtained when titanium compounds were used. Similar results were obtained when MgI_2 was substituted for elementary iodine.

Vol'pin and Shur /96/ suggested a tentative mechanism for the absorption of nitrogen. In their view, molecular nitrogen becomes activated by complex formation with the transition metal hydride or alkyl, which is formed as intermediate product. Such a complex can be a typical π-complex:

$$>Me \leftarrow \begin{matrix} N \\ ||| \\ N \end{matrix}$$

or a metal carbonyl-like complex:

$$>Me^{(-)}-N^{(+)}\equiv N \rightleftarrows \geqslant Me=N^{(+)}=N^{(-)} \rightleftarrows \geqslant Me^{(-)}-N=N^{(+)}$$

In such a complex nitrogen can be reduced to ammonia or may form a compound of the nitride type.

According to these workers, complex formation between nitrogen and the metal ions in a lower valency than in the initial salt is not very likely; the same applies to the possible formation of nitrides with the free metal in the active state, obtained by decomposition of metal alkyl or metal hydride. It must be pointed out that experiments on the nitrogen absorption by systems which do not contain Me — C or Me — H bonds /102/ are in contradiction with this mechanism. It is assumed by Shilov and Nechiporenko /103/, who studied the fixation of nitrogen by the system $(C_5H_5)_2TiCl_2 - C_2H_5MgBr$, that nitrogen reacts with the active metal formed by decomposition of the initially produced hydride, with formation of nitrides as the final product.

Bibliography

1. Ziegler, K., H. Gellert, E. Holzkamp, and G. Wilke. — Brennstoff Chem., **35**:321. 1954; Angew. Chem., **67**:541. 1955.
2. Zhigach, A. F. (editor). Alyuminiiorganicheskie soedineniya (Organoaluminum Compounds). — IL. 1962. [Collection of translated articles].
3. Gaylord, N. G. and H. F. Mark. Linear and Stereoregular Addition Polymers. — New York, Interscience. 1959.
4. Natta, G. — Chem. Ind., p. 223. 1965.
5. Ewers, J. — Angew. Chem., **78**:593. 1966.
6. Ziegler, K. et al. — Lieb. Ann. Chem., **629**:121. 1960; see also /2/, p. 170.
7. Anhorn, V., K. Fresh, et al. — Chem. Eng. Progr., **57**:43. 1961.
8. Baas, C. and J. Vlugter. — Brennstoff Chem., **45**(161):295. 1964.
9. Landau, R., R. Simon, and G. Schaffel. — Chim. ind., **90**:37. 1964.

9a. Fel'dblyum, V. Sh. and N. V. Obeshchalova. — Doklady AN SSSR, **172**:368. 1967.

9b. Fel'dblyum, V. Sh. et al. — Neftekhimiya, **7**:380. 1967.

10. Ziegler, K. — Brennstoff Chem., **45**:194. 1964.
11. Ziegler, K. — Angew. Chem., **68**:721, 729. 1956.
12. Sun Oil Co., US Patent 3035077. 1962; C. A., **57**:11021. 1962.
13. BASF A.G., GFR Patent 1186046. 1965; C. A., **62**:9007. 1965.
14. Natta, G. and A. Miyake. — Bull. Chem. Soc. Japan, **38**:351. 1965.
15. Wilke, G. — Angew. Chem., **69**:397. 1957.
16. Wilke, G. — Angew. Chem., **75**:10. 1963.
17. Takahasi, H. and M. Yamaguchi. — J. Org. Chem., **28**:1409. 1963.
18. Wilke, G. — Angew. Chem., **68**:306. 1956.
19. Müller, H. — GFR Patent 1097982. 1961. GFR Patent 1095819. 1961.
20. GFR Patent 1106758. 1961.
21. Wilke, G. — GFR Patent 1078108. 1958; J. Polymer Sci., **38**:45. 1959.
22. Otsuka, S. et al. — Kogyo Kagaku Zasshi, **66**:1094. 1963.
23. Müller, H., D. Wittenberg, H. Seibt, and E. Scharf. — Angew. Chem., **77**:318. 1965.
24. French Patent 1350644. 1963.
25. Otsuka, S. et al. — J. Am. Chem. Soc., **85**:3709. 1963; Saito, T. et al. — Bull. Chem. Soc. Japan, **37**:105. 1964.
26. Tamai, K., T. Saito, J. Uchida, and A. Misono. — Bull. Chem. Soc. Japan, **38**:1575. 1965.
27. Natta, G. — J. Am. Chem. Soc., **86**:3903. 1964.

27a. Iwamoto, Masao and Sadao Yuguchi. — Japanese Patent 7944, Class 16B123. 1966.

28. Natta, G., G. Mazzanti, et al. — Angew. Chem., **69**:685. 1957; Atti Accad. Lincei, **25**:3. 1958; Gazz., **89**:465. 1959.
29. Berlini, A. A. et al. — VMS, **1**:1817. 1959; **5**:1354. 1963; Izvestiya AN SSSR, OKhN, p. 1875. 1965.
30. Nasirov, F. M., B. A. Krentsel', and B. E. Davydov. — Izvestiya AN SSSR, OKhN, p. 1009. 1965.
31. Lombardi, E. and L. Giuffre. — Atti Accad. Lincei, **25**:70. 1958.
32. Hagihara, N., M. Tamura, H. Yamazaki, and M. Fujiwara. — Bull. Chem. Soc. Japan, **34**:892. 1961.
33. Reikhsfel'd, V. O. and K. L. Makovetskii. — Uspekhi Khimii, **35**:1204. 1966.

34. Franzus, B., P. Canterino, and R. Wicklife. — J. Am. Chem. Soc., **81**:1514. 1959.
35. Lutz, E. — J. Am. Chem. Soc., **83**:2551. 1961.
36. Reikhsfel'd, V.O., K.L. Makovetskii, and L.L. Erokhina. — ZhOKh, **32**:653. 1962.
37. Makovetskii, K.L., V.O. Reikhsfel'd, and L.L. Erokhina. — ZhOKh, **34**:1962. 1964.
38. Makovetskii, K.L., V.O. Reikhsfel'd, and L.L. Erokhina. — Zhurnal Organicheskoi Khimii, **2**:753. 1966.
39. Makovetskii, K.L. Author's Summary of Thesis, LTI im. Lensoveta. 1964; Reikhsfel'd, V.O. and K.L. Makovetskii. — Doklady AN SSSR, **155**:414. 1964; Makovetskii, K.L. et al. — Zhurnal Organicheskoi Khimii, Part 2, p. 759. 1966.
40. Makovetskii, K.L., B.I. Lein, and V.O. Reikhsfel'd. — ZhOKh, **34**:3505. 1964; Reikhsfel'd, V.O., B.I. Lein, and K.L. Makovetskii. — Zhurnal Organicheskoi Khimii, **2**:961. 1966.
41. Hoover, F., O. Webster, and C. Handy. — J. Org. Chem., **26**:2234. 1962.
42. Kambara, S., M. Hatano, N. Sera, and K. Muran. — Khimiya i tekhnologiya polimerov, p. 91. 1963.
43. US Patent 3082269. 1963.
44. Wilke, G. — Angew. Chem., **72**:581. 1960.
45. Prince, M. and K. Weiss. — J. Organomet. Chem., **2**:251. 1964.
46. Chini, P., G. De Venuto, T. Salvatori, and M. De Malde. — Chim. ind. (Milan), **46**:1049. 1961.
47. Luttinger, L. and E. Colthup. — J. Org. Chem., **27**(1591):3752. 1962.
48. Chauvin, V. and G. Lefebre. — Comptes rendus, **259**:2105. 1964; Bull. Soc. chim. France, p. 3223. 1966.
49. GFR Patent 1180362. 1964; C.A., **62**:9004. 1965.
50. Symcox, R.O. — J. Polymer Sci., **B2**:947. 1964.
51. Shimizu, A., T. Otsu, and M. Imoto. — J. Polymer Sci., **B3**:449. 1965.
52. Shimizu, A. et al. — J. Polymer Sci., **B3**:1031. 1965.
53. Aubrey, D.W., A. Barnatt, and W. Gerrard. — J. Polymer Sci., **B3**:357. 1965.
54. Schindler, A. — Makromol. Chem., **90**:284. 1966.
55. Goodrich, J. and R. Porter. — J. Polymer Sci., **B2**:353. 1964.
56. Marvel, C. and J. Rogers. — J. Polymer Sci., **49**:333. 1961.
57. Ermakova, I.I. Author's Summary of Thesis, INKhS AN SSSR, Moskva. 1965.
58. Golub, M. — Chem. Eng. News, **15**:44. 1962; Canad. J. Chem., **41**:937. 1963.
59. Dolgoplosk, B.A., E.N. Kropacheva, and K.V. Nel'son. — Doklady AN SSSR, **123**:685. 1958.
60. Boldyreva, I.I., B.A. Dolgoplosk, E.N. Kropacheva, and K.V. Nel'son. — Doklady AN SSSR, **131**:830. 1958.
61. Shelton, C. and L. Lee. — Rubb. Chem. Techn., **31**:415. 1958.
62. Yuguchi, S. and M. Iwamoto. — J. Polymer Sci., **B2**:1035. 1964.
63. Gol'dfarb, Yu. Ya. Author's Summary of Thesis, INKhS AN SSSR, Moskva. 1964.
64. Korneev, N.N., S.K. Goryunovich, and I.F. Leshcheva. — Plasticheskie Massy, No. 1:10. 1968.

65. Groizeleau, L. — Comptes rendus, **244**:1223. 1957.
66. Nicolescu, I. V., M. Iovu, and G. I. Nikishin. — Izvestiya AN SSSR, OKhN, No. 1:94. 1960.
67. British Patent 997303. 1965; C. A., **63**:13145. 1965.
68. Dutch Patent 6400343. 1964; C. A., **62**:7679. 1965.
69. Nicolescu, I. V. and O. Serban. — Stud. cercet., **13**:525. 1965.
70. Gaylord, N. G. Reduction with Complex Metal Hydrides. — New York, Interscience. 1956.
71. Takegami, Y., T. Ueno, and T. Fujii. — Bull. Chem. Soc. Japan, **38**:1279. 1965.
72. Takegami, Y., T. Ueno, and T. Fujii. — Kogyo Kagaku Zasshi, **69**:1467. 1966.
73. Chem. Eng. News, p. 34. 1963.
74. US Patent 3205278. 1965; Organomet. Compds., **7**:161. 1965.
75. Sloan, M., A. Matlack, and D. Breslow. — J. Am. Chem. Soc., **85**:4014. 1963.
76. Kalechits, I. V. and F. K. Shmidt. — Kinetika i Kataliz, **7**:614. 1966.
76a. Lipovich, V. G., F. K. Shmidt, and I. V. Kalechits. — Kinetika i Kataliz, **8**:1300. 1967.
77. Zakharkin, L. I., V. V. Gavrilenko, and O. Yu. Okhlobystin. — Izvestiya AN SSSR, OKhN, p. 100. 1958.
78. Podall, H. — J. Am. Chem. Soc., **80**:5573. 1958.
79. Podall, H. and H. Shapiro. — Abstracts of the 136th Meeting Am. Chem. Soc., 51N, p. 112. 1959.
80. Podall, H., J. Dunn, and H. Shapiro. — J. Am. Chem. Soc., **82**:1325. 1960.
81. US Patents 2952517, 2952518. 1960; C. A., **55**:4903. 1961.
82. US Patents 2952212—2952214. 1960; C. A., **55**:4903. 1961.
83. Podall, H. and A. Giraitis. — J. Org. Chem., **26**:2587. 1961.
84. GFR Patent 113688. 1961; C. A., **56**:8297. 1962.
85. Szabó, P. and L. Markó. — J. Organomet. Chem., **3**:364. 1965.
86. Bennett, M. A. — Chem. Revs, **62**:611. 1962.
87. French Patent 1320729. 1963; C. A., **59**:14026. 1963.
88. GFR Patent 1191375. 1965; Organomet. Compds., **6**:253. 1965.
89. Schrauzer, G. N. — In: Advances in Organometallic Chemistry, **2**, New York, Academic Press. 1964.
90. Wilke, G. and G. Herrmann. — Angew. Chem., **74**:693. 1962.
91. Tsutsui, M. and G. Chang. — Canad. J. Chem., **41**:1255. 1963.
92. Tsutsui, M. and M. Levy. — Proc. Chem. Soc., p. 117. 1963.
93. Yamomoto, H. et al. — J. Am. Chem. Soc., **87**:4652. 1965.
94. Wilke, G. and G. Herrmann. — Angew. Chem., **78**:591. 1966.
95. Wilke, G. and H. Scott. — Angew. Chem., **78**:592. 1966.
96. Vol'pin, M. E. and V. B. Shur. — Vestnik AN SSSR, p. 51. 1965.
97. Vol'pin, M. E. and V. B. Shur. — Doklady AN SSSR, **156**:1102. 1964.
98. Vol'pin, M. E., V. B. Shur, and L. P. Bichin. — Izvestiya AN SSSR, OKhN, p. 720. 1965.
99. Vol'pin, M. E., N. K. Chanovskaya, and V. B. Shur. — Izvestiya AN SSSR, OKhN, p. 1082. 1966.
100. Vol'pin, M. E., M. A. Ilatovskaya, E. I. Larikov, M. L. Khidekel', Yu. A. Shvetsov, and V. B. Shur. — Doklady AN SSSR, **164**:331. 1965.

101. Khidekel', M.L. and Yu.B. Grebenshchikov. — Izvestiya AN SSSR, p.761. 1965.
102. Vol'pin, M.E., A.A. Belyi, and V.B. Shur. — Izvestiya AN SSSR, OKhN, p.2225. 1965.
103. Nechiporenko, G.N., G.M. Tabrina, A.K. Shilova, and A.E. Shilov. — Doklady AN SSSR, **164**:1062. 1965.

Chapter VII

TECHNIQUES EMPLOYED IN WORKING WITH COMPONENTS OF CATALYST COMPLEXES

The components of catalyst complexes used in various branches of chemistry are highly reactive towards different compounds, including water and oxygen.

Titanium tetrachloride and vanadium tetrachloride are the transition metal salts most often employed. They fume strongly in the air, owing to the reaction with atmospheric moisture, with formation of hydrogen chloride. As a result, they become less active in polymerization processes. Trivalent titanium and vanadium salts are also oxidized in the air to the tetravalent state, which also strongly affects their catalytic activity in the complex. Organoaluminum compounds are much more reactive than transition metal salts. They spontaneously catch fire in the air, and explode in contact with water, the explosion being accompanied by self-ignition.

Only fragmentary literature information /1 — 7/ is so far available on safe working techniques, first aid and antifire precautions to be taken during the industrial manufacture of these compounds; the detailed review by Knap et al. /1/ is the only exception.

In this chapter an attempt is made to summarize all the information available on the subject for the benefit of industrial undertakings and research laboratories.

1. CERTAIN CHARACTERISTIC PROPERTIES OF ALKYLALUMINUM COMPOUNDS

Alkylaluminum compounds display a strong affinity to oxygen and water; this particularly applies to trimethylaluminum, triethylaluminum, diethylaluminum hydride, diethylaluminum chloride, triisobutylaluminum and diisobutylaluminum chloride, which are very dangerous in use. Table 18 represents the results obtained by Baratov et al. for the flash points and limiting detonation concentrations of alkylaluminum compounds in the air. It is seen from the table that these products are spontaneously flammable at very low temperatures, except for diisobutylaluminum chloride /8/. It was found that if alkylaluminum compounds are diluted with hydrocarbons, their flash points become higher. Thus, for instance, a 40% solution of triethylaluminum in gasoline undergoes spontaneous combustion only at room temperature, while a 15% solution does not undergo spontaneous combustion at all. Solutions of crude trialkylaluminum compounds which contain finely dispersed active aluminum also catch fire. The danger of spontaneous combustion of lower alkylaluminum compounds and dialkylaluminum halides is further intensified by possible exothermal side reactions which result in the liberation of gaseous products.

Another reason for the great danger of explosion, produced by spontaneous combustion or even simply by heating (especially as regards triisobutylaluminum), is that alkylaluminum compounds decompose on being heated with formation of olefins. The decomposition of triisobutylaluminum begins at a temperature as low as 50°C, and the decomposition rate rapidly increases with the temperature. As a result, the concentration of the gases per unit volume increases and the detonation limit is attained.

We have already mentioned that one of the most dangerous reactions is that between alkylaluminum compounds and oxygen. This reaction is highly exothermal and takes place spontaneously when the substances are brought into contact. If alkylaluminum compounds are stored in closed vessels, any oxidation will be accompanied by thermal decomposition, with the result that a high pressure will build up and the vessel will be likely to burst. The heat evolved during such an oxidation is comparable to the heat of combustion of octane /1/.

In our view, the highly dangerous reactions between alkylaluminum compounds and oxygen still involve a smaller risk than working with vessels and apparatus which contain even traces of water. In working with alkylaluminum compounds, water will be found to be the most active substance, obviously except for special oxidizing agents such as H_2O_2, etc. Such reactions are instantaneous, highly exothermal, explosive, and are accompanied by spontaneous combustion. Alcohols, acids, primary and secondary amines react violently with alkylaluminum compounds, but less so than water. These compounds can be arranged in the following series of reactivity with respect to alkylaluminum compounds:

$$H_2O > C_2H_5OH > C_8H_{17}OH > R_2NH$$

TABLE 18. Flash points and spontaneous combustion range of the most important alkylaluminum compounds

Compound	Flash point, °C	Flammability range, vol.%	
		lower	upper
Triethylaluminum	−68	1.88	13.1
Triisobutylaluminum	−40	1.53	8.7
Diethylaluminum chloride	−60	2.17	12.1
Diisobutylaluminum chloride	2	1.4	8.25

It is not recommended, accordingly, to use these compounds as heating agents in the preparation of alkylaluminum compounds. If an alkylaluminum compound is allowed to leak into the atmosphere on plant premises, white, musty-smelling fumes appear. These fumes are also formed if dilute solutions of these compounds are oxidized in the air. It was shown by Sanotskii /9/ that the reaction between alkylaluminum compounds and the atmosphere yields a complex mixture of oxidation and hydrolysis products; its major components are aluminum aerosol, alumina, hydrogen chloride (in the case of alkylaluminum chlorides) and unsaturated hydrocarbons.

Subsequent studies confirmed these results; moreover, a number of other compounds were identified in the mixture /10/, including various alcohols and aldehydes, carbon monoxide and alkyl halides (in the case of decomposition of alkylaluminum chlorides). It will be noted that the concentration of the alumina aerosol increases owing to the settling of the particles. It was also found that most of the hydrolysis products in the atmosphere are not found free, but are adsorbed on aerosols /9/.

Triethylaluminum, diethylaluminum chloride and diisobutylaluminum chloride, present in respective concentrations of 10, 11 and 20 mg/liter of aerosol proved 100% lethal to white rats; they all proved 50 % lethal at about 7 mg/liter; the absolutely transferable concentrations are 2 — 3 mg/liter. It was also found that the toxicity of the compounds varies with atmospheric humidity. Advanced poisoning is accompanied by irritation of the mucous membranes of the eye and upper respiratory ducts, and by the depression of the central nervous system. The consumption of oxygen decreases at the same time. The dead rats displayed plethora of the brain and of the internal organs and also acute emphysema and small subpleural hemorrhages. It was established that the maximum permissible concentration of trialkylaluminum compounds and alkylaluminum chlorides in the atmosphere on plant premises is about 0.0007 mg/liter (aerosol) and 0.0001 mg/liter (hydrogen chloride).

Concentrated alkylaluminum compounds which have come into contact with the human skin at first cause intense pain, which then gives way to a strong burning sensation on the skin; this eventually disappears. Single blisters then form on the place of contact, their size varying from that of a bean to that of a child's fist. Anton'ev and Raben /11/ studied the origin of the burns produced by alkylaluminum compounds and the treatment of such burns. They found that if a secondary pyococcal infection is present, the contents of the blisters are turbid, and the blister itself is surrounded by a bright-red inflamed ring, 2 — 3 mm wide. The blisters are easily ruptured by external traumas and may even burst spontaneously. The sore heals very slowly, but no scars are left. The burned areas become covered by thin, pinkish epidermis, which may subsequently assume secondary hyperpigmentation.

The treatment consists in aseptic opening of the blister and in twice-daily application of linol* dressings. This treatment was successfully applied by Dolgov et al. /12/ in the treatment of radiation burns on the skin.

A study of the effect of organoaluminum compounds on the main brands of construction steels showed that practically all of them are suitable for the construction of apparatus, ducts and installations used in the manufacture of these compounds. They are also inert toward tin and silver solders. The most suitable lining materials are copper and Teflon. Flexible polyethylene hoses are very suitable for working with trialkylaluminum compounds; these hoses are unsuitable for working with alkylaluminum halides. Rubber hoses may be used only if the time of contact with the products is limited /1/. All fabrics are attacked by alkylaluminum compounds, even in an inert gas medium.

[* Mixture of methyl esters of fatty acids found in linseed oil.]

2. WORKING TECHNIQUES

Owing to the specific properties of organoaluminum compounds just described, working with them requires a special technique, both in the laboratory and during production on an industrial scale.

All contact of alkylaluminum compounds with oxygen and water must be avoided. This is achieved by flushing all apparatus, installations and vessels with dry nitrogen or argon, purified from oxygen. The flushing is continued until the last traces of atmospheric oxygen and water have been removed; the permissible residual content of oxygen is 0.01 vol. %, that of water not more than 0.01%.

Experimental work is most often conducted in airtight metallic apparatus. If glass vessels must be employed, they must be made of thick-walled, heat-resistant glass, and the amounts of alkylaluminum compounds involved must not exceed 50 g.

Working with alkylaluminum compounds in special apparatus of different designs has been described in detail /13, 14/. A very useful technique is to work in special airtight nitrogen chambers. In its simplest form, the nitrogen chamber consists of a box made of organic glass, permanently sealed to a metallic bottom. One or two walls of the box are fitted with holes for rubber gloves to be used when handling the materials or the equipment inside the box. One of the side walls contains a lock chamber, through which materials may be taken in or out of the chamber. Special ducts leading into the lock chamber and the main chamber serve to pass a stream of nitrogen free from all traces of moisture. The chamber must be airtight; it is usually kept under excess pressure of nitrogen in order to prevent penetration of air.

Small amounts of alkylaluminum compounds may be destroyed by combustion in a special place reserved for the purpose, after previous dilution with petroleum or mineral oil; if such a place is not available, the solution may be decomposed by any higher alcohol at a low temperature. When the decomposition is complete, some water is added and the reaction mixture is heated to complete the decomposition. It is recommended that large amounts (more than one liter) of alkylaluminum compounds be burned.

Industrial plants engaged in the manufacture of organoaluminum compounds must have perfectly airtight equipment and installations. The electrical leads must be screened in order to ensure airtight, explosion-proof installations /15/.

Storage and transport of organoaluminum compounds must follow the procedures prescribed for easily flammable substances /17/. They should be stored and transported in steel tanks or other metal vessels, which have been made airtight by means of valves. Before filling, the metal vessel should be carefully cleaned of all extraneous matter, dried and filled with dry nitrogen to an excess pressure of 3 — 5 atm. Only then is it permissible to proceed with the filling of the vessel with the organoaluminum compounds, through dry pipes, flushed with dry oxygen-free nitrogen. It is recommended to maintain a small overpressure (about 0.1 atm) of nitrogen in vessels employed for the storage of organoaluminum compounds, in order to prevent penetration of air.

If a small sample of the compound is to be withdrawn from metallic vessels into glass receivers for analysis or research purposes, the receivers should be airtight, and should be provided with airtight stopcocks lubricated with grease which is resistant to organoaluminum compounds. The glass receivers and the tubes interconnecting the vessels should be also carefully flushed with purified, dry nitrogen, passed through a route including the entire space inside the vessel /17/.

3. SAFETY MEASURES, FIRST AID AND FIRE PRECAUTIONS

If the measures just described are adopted, the products cannot leak outside the installation and the danger of fire and to the health of the workers will be reduced to a minimum. However, when working with alkylaluminum compounds one must always be prepared for the possibility of their catching fire at any moment. Accordingly, the work must be carried out very cautiously and deliberately.

All precautions notwithstanding, accidents may always happen. It was pointed out by Anton'ev and Raben /11/ that these occur most frequently during attempts to transfer small amounts of substances from one vessel to another with disregard of the safety rules. Such accidents most often result in burns. It is accordingly necessary to work behind protective screens made of organic glass. When pouring liquid products, protective goggles (or baffle visors made of organic glass) must be used, as well as leather gloves to protect the skin. Personnel in industrial plants producing organoaluminum compounds should wear protective clothing; such clothing is best made of so-called metallized fabrics or leather. This is a fabric or leather cloth coated with a thin layer of aluminum and aluminum oxide powder. If a small amount of an alkylaluminum compound has been inadvertently touched, the affected place should be washed with a large amount of water or kerosine, rubbed with alcohol and dressed with linol, supplies of which should always be available in the first aid box or first aid station of the laboratory or of the industrial plant. If the burn is extensive (i.e., if a large surface is affected), the affected part should be washed with a stream of water (best of all in a special shower room), after which medical assistance should be sought.

If a solution of an alkylaluminum compound has been spilled and large concentrations of white fumes are evolved as a result, gas masks with an anti-fume filter must be put on; if it is necessary to stay in the fire zone for a longer time, insulating oxygen counter-gases should be released.

Since alkylaluminum compounds must be produced on a large scale which constitutes a considerable fire hazard, attempts have been made in many countries to develop suitable procedures for fighting these fires. It was found that a number of conventional fire-fighting agents do not give satisfactory results with alkylaluminum compounds. Water and aqueous fire-fighting agents are unsuitable, since they react with the compounds. Carbon dioxide is suitable for quenching small fires, but not if the fire is extensive /8/. Dry chemical fire extinguishers and chlorobromomethane are employed against fires in thin layers and in the bulk. These substances, while more effective than carbon dioxide, are incapable of dealing with the fire completely and of preventing the fire from spreading /18/. Thus, the important consideration is not so much the choice of a suitable chemical, as the choice of an agent which would absorb the residual matter and thus prevent the fire from spreading.

Tseratskii, who studied spontaneous combustion and fire-fighting techniques when working with alkylaluminum compounds, proposed a fire-fighting substance, consisting of a dry chemical containing a bicarbonate base and a dry, finely ground graphite sorbent, which can be conveniently released from dry chemical extinguishers /18/. The technique resembles that employed in extinguishing liquid fires by means of an ordinary dry extinguishing agent. Tseratskii's fire-fighting substance acts as follows: in the first stage, the flame is extinguished, after which the residual alkylaluminum compound is absorbed, so that the fire is prevented from spreading. The organoaluminum compound undergoes a slight decomposition on being brought into contact with the fire-fighting substance, but this does not interfere with the extinction of the fire. The ratio of the extinguishing agent to any single alkylaluminum compound varies between 8:1 for fire-fighting in the bulk, and 10:1 for extinguishing a thin layer fire.

Baratov et al. suggested that the most effective fire-fighting agent available — SI-2 — be used for fighting alkylaluminum fires. The expenditure of the SI-2 composition in extinguishing burning organoaluminum compounds in a concentrated form, in layers not more than 10 cm thick, is 1 kg composition per 2 kg of the alkylaluminum compound. The agent SI-2 may be applied from special fire-extinguishers or else from stationary and mobile installations /8/. Another suitable fire-extinguishing agent is sodium chloride, treated with a 0.05% aqueous solution of fuchsine. Burning dilute solutions of alkylaluminum compounds (not more concentrated than 10%) may be extinguished with mechanical air foam and with a sheet of finely dispersed water droplets.

Bibliography

1. Knap, J. et al. — Ind. Eng. Chem., **49**:874. 1957.
2. Nobis, J. — Ind. Eng. Chem., **49**(12):44A. 1957.
3. Hamprecht, G. and H. Muehlbauer. — GFR Patent 1004179. 1957.
4. Chem. Eng. News, **34**:4826. 1956.
5. Chem. Age, **77**:963. 1957.
6. Chem. Labor. u. Betr., **5**:216. 1958.
7. Ind. Eng. Chem., **50**(7):77A. 1958.
8. Ryabov, I. V. (editor). Pozharnaya opasnost' veshchestv i materialov (Fire Hazard of Substances and Materials) (Handbook). — Stroiizdat. 1966.
9. Sanotskii, I. V. Materialy nauchnoi sessii po toksikologii vysokomolekulyarnykh soedinenii (Data of the Scientific Session on Toxicology of High-Molecular Compounds), Moskva-Leningrad, p. 50. 1960.
10. Krivoruchko, F. D. — Gigiena i Sanitariya, **8**:57. 1966.
11. Anton'ev, A. A. and A. S. Raben. — Gigiena Truda i Profzabolevanii, **5**:51. 1965.
12. Dolgov, A. P. et al. — Trudy V Vsesoyuznogo s''ezda dermatologov, Leningrad, p. 249. 1961.
13. Kocheshkov, K. A. and A. N. Nesmeyanov. Sinteticheskie metody v oblasti metallorganicheskikh soedinenii 3 gruppy (Methods for the Synthesis of Organic Compounds of Group III Metals). — Izd. AN SSSR. 1945.

14. Nesmeyanov, A.N. and R.A.Sokolik. Metody elementoorganicheskoi khimii (Methods of Heteroorganic Chemistry). — Izd. "Nauka." 1964.
15. Chernousov, N.P., A.N.Kutin, and V.F.Fedorov. Germeticheskie khimikotekhnologicheskie mashiny i apparaty (Hermetic Chemical Processing Machines and Devices). — Izd. "Mashinostroenie." 1965.
16. Kostin, N.V. Tekhnika bezopasnosti raboty v khimicheskikh laboratoriyakh (Prevention of Work Accidents in Chemical Laboratories). — Izd. MGU. 1966.
17. Zhigach, A.F. and D.S.Stasinevich. — Reaktsii i metody issledovaniya organicheskikh soedinenii, No. 10:177. 1961.
18. Zeratsky, E. — Hydrocarbon Process. Petrol. Refiner, 40(8):145. 1961.

LIST OF ABBREVIATIONS APPEARING IN THE TEXT

AN KazSSR	Akademiya Nauk Kazakhskoi SSR	Academy of Sciences of the Kazakh SSR
AN SSSR	Akademiya Nauk SSSR	Academy of Sciences of the USSR
F KhI	Fiziko-Khimicheskii Institut	Institute of Physical Chemistry
IL	Izdatel'stvo Inostrannoi Literatury	Foreign Literature Publishing House
IN KhS AN SSSR	Institut Neftekhimicheskogo Sinteza im. A. V. Topchieva Akademii Nauk SSSR	Topchiev Institute of Petrochemical Synthesis of the Academy of Sciences of the USSR
LTI	Leningradskii Tekhnologicheskii Institut imeni Lensoveta	Lensovet Leningrad Technological Institute
MGU	Moskovskii Gosudarstvennyi Universitet	Moscow State University
NIISS	Nauchno-Issledovatel'skii Institut Sinteticheskikh Spirtov i Organicheskikh Produktov	Scientific Research Institute of Synthetic Alcohols and Organic Products
NIITEKhIM	Nauchno-Issledovatel'skii Institut Tekhniko-Ekonomicheskikh Issledovanii po Khimii	Scientific Research Institute for Technical and Economic Investigations in Chemistry
OKhN	Otdelenie Khimicheskikh Nauk (Akademii Nauk SSSR)	Department of Chemical Sciences (of the Academy of Sciences of the USSR)
VIMS	Vsesoyuznyi Nauchno-Issledovatel'skii Institut Mineral'nogo Syr'ya	All-Union Scientific Research Institute for Raw Minerals
VMS	Vysokomolekulyarnye Soedineniya	High-Molecular Compounds (Journal)
ZhAKh	Zhurnal Analiticheskoi Khimii	Journal of Analytical Chemistry
ZhFKh	Zhurnal Fizicheskoi Khimii	Journal of Physical Chemistry

ZhNKh	Zhurnal Neorganicheskoi Khimii	Journal of Inorganic Chemistry
ZhOKh	Zhurnal Obshchei Khimíi	Journal of General Chemistry
ZhPKh	Zhurnal Prikladnoi Khimii	Journal of Applied Chemistry